AF875951

Matthias Micha Wildermuth

Quantum Tunneling of Josephson Vortices in High-Impedance Long Junctions

Experimental Condensed Matter Physics
Band 33

Herausgeber
Physikalisches Institut
Prof. Dr. David Hunger
Prof. Dr. Alexey Ustinov
TT-Prof. Dr. Philip Willke
Prof. Dr. Wolfgang Wernsdorfer
Prof. Dr. Wulf Wulfhekel

Eine Übersicht aller bisher in dieser Schriftenreihe erschienenen Bände finden Sie am Ende des Buchs.

Quantum Tunneling of Josephson Vortices in High-Impedance Long Junctions

by

Matthias Micha Wildermuth

Scientific Publishing

Karlsruher Institut für Technologie
Physikalisches Institut

Quantum Tunneling of Josephson Vortices in
High-Impedance Long Junctions

Zur Erlangung des akademischen Grades eines Doktors der Naturwissenschaften von der KIT-Fakultät für Physik des Karlsruher Instituts für Technologie (KIT) genehmigte Dissertation

von Matthias Micha Wildermuth, M.Sc.

Tag der mündlichen Prüfung: 22. Juli 2022
Referent: Prof. Dr. Alexey V. Ustinov
Korreferent: Prof. Dr. Alexander Shnirman

Data availability: https://doi.org/10.1063/5.0082197

Impressum

KIT Scientific Publishing

Karlsruher Institut für Technologie (KIT)
KIT Scientific Publishing
Straße am Forum 2
D-76131 Karlsruhe

KIT Scientific Publishing is a registered trademark
of Karlsruhe Institute of Technology.
Reprint using the book cover is not allowed.

www.ksp.kit.edu

This document – excluding parts marked otherwise, the cover, pictures and graphs – is licensed under a Creative Commons Attribution-Share Alike 4.0 International License (CC BY-SA 4.0): https://creativecommons.org/licenses/by-sa/4.0/deed.en

The cover page is licensed under a Creative Commons Attribution-No Derivatives 4.0 International License (CC BY-ND 4.0): https://creativecommons.org/licenses/by-nd/4.0/deed.en

Print on Demand 2023 – Gedruckt auf FSC-zertifiziertem Papier

ISSN 2191-9925
ISBN 978-3-7315-1303-2
DOI 10.5445/KSP/1000158887

Contents

1 Introduction

The occurrence of vortices is a widespread phenomenon in physics. They are well known, for instance, from fluid mechanics of gases and liquids [Bes06; ZB22], but vortices also occur at cryogenic temperatures in superfluids and in the condensate of superconducting charge carriers in solids [EH05; Tin04]. In the latter case, vortices of supercurrents exist in Josephson junctions that are superconducting leads separated by a tunnel barrier [Jos62; AR63]. Here, if the junction is tailored to an appropriate shape, it allows for spatial and temporal phase distribution with changes in integer multiples of 2π [Jos64; Swi61]. This phenomenon gives rise due to a Bohr-Sommerfeld-like quantization of the superconductor's phase and is analogous to the angular momentum quantization in atoms [Lon50; DN61; DF61]. As a characteristic of superconductors, such phase drops drive supercurrents, which are closed rings, namely Josephson vortices. They enclose a quantized magnetic flux of $\Phi_0 = h/2e$ [BL64; Jos65], often referred to as "fluxon" [BP82; Lik86].

The basic understanding of the phase dynamics of these vortices comes from the sine-Gordon model [Kul65; Sco69; FD73], which features a nonlinear wave equation for the Josephson medium. Its vortex solution is of particular interest [Ust98], as these excitations are pinned at the tunnel barrier and may propagate along the nonlinear Josephson medium as a wave packet that maintains its shape, i.e., as a soliton [Swi61]. Fluxons in long Josephson junctions have been a rich research field for decades since they serve as an excellent physical system to study highly nonlinear electrodynamics and soliton dynamics on a microscopic scale [Par78; Ped83]. Thereby, classical effects such as high-frequency cavity resonances [CFL71; FD73; FD74b; Fis64; Kul65; Kul67], but also relativistic mass increase and Lorentz contraction [MS78; Lau+95] have been studied. Among others, applications for microwave generation [Nag+83; Kos+01] and amplification [Nag+85; Nor95] based on flux-flow oscillators have been developed.

Although Josephson vortices host quanta of magnetic flux, only their phase is quantized. However, they do not behave like quantum particles because both phase and charge of the vortex commute and hence do not feature Heisenberg's uncertainty relation.

The key question is whether and how the Josephson medium can be modified so that Josephson vortices behave like quantum particles.

Experimentally realized Josephson quantum particles are interesting, as there are theoretical proposals to explore relativistic quantum sine-Gordon physics [Roy+21], and quantum coherent excitations in spatially periodic potentials simulating solid-state models. Moreover, topological solitons would theoretically be auspicious candidates for superconducting quantum bits [KI96; Wal+00b; KWU02; FU03; SK05; KD11] with good energy relaxation and coherence properties. These characteristic time scales are expected to be in the order of hundreds of microseconds [KDP06] since Josephson vortices are well decoupled from other electromagnetic excitations. This weak interaction is due to the vortices' small magnetic dipole moment and the large capacitance of their hosting long junction that serves as an intrinsic filter for low-frequency charge noise below the Josephson plasma frequency. Unlike established superconducting quantum bits, which are all compositions of Josephson junctions, capacitors, and inductors [VD17; Kra+19], these elements are intrinsically included in Josephson media, enabling very compact circuits.

All these considerations are theoretical, and the experimental implementation requires junctions that allow for quantum vortices. The system's quantumness is governed in essence by the junction impedance, i.e., depends on the inductance of the electrodes, the capacitance of the junction, and its width. So far, entering the quantum regime of Josephson vortices has been experimentally very difficult to reach [Wal+03b] and the coherent quantum regime of Josephson vortices has not been realized yet.

The goal of this thesis is the experimental realization of Josephson vortices with a quantum nature in a distributed junction. To increase the long junction's quantumness, one needs to enhance its impedance. We use the approach of very narrow junctions and replace the junction's usual bulk electrodes with thin films of a high-kinetic inductance superconductor. With this additional parameter of kinetic inductance, which exceeds the magnetic contribution by orders of magnitude, the impedance can be increased beyond the purely geometrical limit.

The thesis continues this introduction with a theoretical background on the principles of quantum electrodynamics in superconducting circuits, with particular attention to long Josephson junctions. Following this, experimental methods for cooling to cryogenic temperatures, the measurement setups, and the fabrication techniques developed for impedance-tailored long Josephson junctions are specified.

In the first part of the experimental results, these high-impedance long junctions are characterized by transport measurements to observe Josephson vortices in the presence of high-kinetic inductance electrodes and to check whether the impedance is increased as intended. In the second results chapter, we investigate a suitable manipulation and readout scheme to explore quantum coherence, where a classical long junction is embedded in a microwave resonator. In the third experimental chapter, we experimentally study whether we can indeed observe Josephson vortices with properties of quantum particles with the chosen approach. Finally, I summarize the main results and conclude with possible future experiments that could follow this work.

2 The principles of quantum electrodynamics in long Josephson junctions

This chapter summarizes the theoretical background that supports the understanding of later presented experiments on high-impedance long Josephson junctions (LJJs). As these tunnel junctions are weak link superconductors, basics of the superconducting phase are introduced, where I devote special attention to the kinetic inductance of circuits composed of granular superconductors that is utilized to increase the junction's impedance. After expounding the fundamentals of lumped element Josephson junctions that remain valid for distributed versions, one-dimensional long junctions are elaborated, as experimentally studied throughout this work. Moreover, superconducting microwave circuits are treated, used as a novel tool to read out long junctions.

2.1 Basics of superconductivity

The successful liquefaction of helium via the Hampson–Linde cycle by Heike Kamerlingh Onnes in 1908 [Kam08] opened the research field of low temperatures solid state physics. This technology paved the way for studying metals' electrical resistivity near zero temperature. In 1911, Kamerlingh Onnes discovered the resistance of mercury to drop abruptly to zero below a transition temperature, an effect he named "superconductivity" [Kam11]. Except for exceeding this critical temperature T_c, superconductivity also vanishes above a critical magnetic field H_c, which is also linked with a critical current I_c, generating an intrinsic magnetic field (Silsbee's rule).

Among many pure metals, superconductivity occurs also in certain alloys, intermetallic compounds, and doped semiconductors and the material specific critical temperatures T_c range from below 1 mK for elementary rhodium over 1.2 K and 9.2 K for aluminum and niobium, respectively, up to 139 K for the cuprate $Hg_{0.8}Tl_{0.2}Ba_2Ca_2Cu_3O_{8.33}$[1] [Kit05].

[1] The given T_c values are no physical limit, but merely the extremes yet known at atmospheric pressure.

Beyond the superconductor's zero DC resistance, Walther Meissner and Robert Ochsenfeld found an even more fundamental property: Up to a critical magnetic field, superconductors expel external magnetic fields from their interior regardless of whether the magnetic field is applied before or after the transition to superconductivity [MO33]. This behavior is independent of the path that leads to it, consequently distinguishing superconductors essentially from ideal conductors and characterizing them as ideal diamagnets. Explaining these results was elementary for subsequent phenomenological theories, such as the London theory and the Ginzburg-Landau theory, and suggested already to explain superconductivity as a macroscopic quantum effect.

2.1.1 Phenomenological description

For a phenomenological description of the superconductor's unique electromagnetic behavior of zero DC resistance and ideal diamagnetism, the brothers Fritz and Heinz London [LL35] extended the Drude model of normal metals to superconductors within the scope of classical electrodynamics. London's theory considers the superconducting state at finite temperatures, consisting of two separate, noninteracting fluids, the normal (index $j = \mathrm{n}$) and superconducting (index $j = \mathrm{s}$) charge carriers with either mass m_j, charge q_j and mean scattering time $t_{0,j}$. Electrodynamics of both fluids is described by the Drude model each

$$q_j \mathbf{E} = m_j \left(\frac{\mathrm{d}}{\mathrm{d}t} + \frac{1}{t_{0,j}} \right) \mathbf{v}_j = \frac{m_j}{n_j q_j} \left(\frac{\mathrm{d}}{\mathrm{d}t} + \frac{1}{t_{0,j}} \right) \mathbf{j}_j \tag{2.1}$$

and the normal conducting static case results in Ohm's law $\mathbf{E} = \sigma_\mathrm{n} \mathbf{j}_\mathrm{n}$ with the normal conductivity $\sigma_\mathrm{n} = \frac{n_\mathrm{n} q_\mathrm{n}^2 t_{0,\mathrm{n}}}{m_\mathrm{n}}$. The superconducting state, however, is characterized by zero DC resistance, and thus $t_{0,\mathrm{s}} \to \infty$, which is why, the classical Ohm's law is replaced by the first London equation

$$\frac{\partial \mathbf{j}_\mathrm{s}}{\partial t} = \frac{n_\mathrm{s} q_\mathrm{s}^2}{m_\mathrm{s}} \mathbf{E}. \tag{2.2a}$$

Equation (2.2a) is Newton's second law for superconducting charge carriers and means that electric fields vanish from the superconductor's interior in the stationary state. To examine the magnetic behavior, this first London equation (2.2a) is inserted in Faraday's law $\nabla \times \mathbf{E} = -\frac{\partial \mathbf{B}}{\partial t}$ and by taking magnetic flux expulsion from the inside, according to the Meissner-Ochsenfeld effect, into account, the second London equation is found to be

$$\nabla \times \mathbf{j} = -\frac{n_\mathrm{s} q_\mathrm{s}^2}{m_\mathrm{s}} \mathbf{B}. \tag{2.2b}$$

These two London equations (2.2a) and (2.2b) can be combined using the electromagnetic vector potential $\mathbf{A}$ and the London gauge and result in

$$\mathbf{j} = \frac{n_\mathrm{s} q_\mathrm{s} \hbar}{m_\mathrm{s}} \boldsymbol{\nabla}\theta - \frac{n_\mathrm{s} q_\mathrm{s}^2}{m_\mathrm{s}} \mathbf{A}. \tag{2.2c}$$

The additional term ($\propto \boldsymbol{\nabla}\theta$) in Eq. (2.2c) originates from London's later derivation [Lon48] that is based on the postulate of superconductivity as a macroscopic quantum state, described by one single wave function encompassing all superconducting charge carriers

$$\psi(\mathbf{r}) = \psi_0(\mathbf{r}) \mathrm{e}^{\mathrm{i}\theta(\mathbf{r})} = \sqrt{n_\mathrm{s}(\mathbf{r})} \mathrm{e}^{\mathrm{i}\theta(\mathbf{r})}. \tag{2.3}$$

The Meissner-Ochsenfeld effect is comprehended by these basic assumptions: By applying the Maxwell equations to the London equation (2.2c), the magnetic flux density and current density have to satisfy the differential equations

$$\nabla^2 \mathbf{B} = \mu_0 \frac{n_\mathrm{s} q_\mathrm{s}^2}{m_\mathrm{s}} \mathbf{B}, \tag{2.4a}$$

$$\nabla^2 \mathbf{j} = \mu_0 \frac{n_\mathrm{s} q_\mathrm{s}^2}{m_\mathrm{s}} \mathbf{j}, \tag{2.4b}$$

respectively. Within a bulk superconductor, the solution of this second-order differential equation describes screening currents underneath the superconductor's surface that decay exponential on the length scale of the so-called London penetration depth

$$\lambda_\mathrm{L} = \sqrt{\frac{m_\mathrm{s}}{\mu_0 n_\mathrm{s} q_\mathrm{s}^2}}. \tag{2.5}$$

In the interior of the bulk superconductor, these screening currents compensate an applied magnetic field H_a, manifesting ideal diamagnetism with $\mu_\mathrm{r} = 0$.

Apart from explaining zero direct current (DC) resistance and the Meissner-Ochsenfeld effect of superconductors, the London theory can explain the superconductor's response to alternating (AC) fields $\mathbf{E}_\mathrm{ac} = \mathbf{E}\mathrm{e}^{\mathrm{i}\omega t}$ and consequently $\mathbf{j}_\mathrm{ac} = \mathbf{j}\mathrm{e}^{\mathrm{i}\omega t}$. Within the scope of the two-fluid model, the normal conducting and the superconducting charge carriers exhibit a complex net conductivity [Tin04]

$$\mathbf{j} = \mathbf{j}_\mathrm{n} + \mathbf{j}_\mathrm{s} = \left(\frac{n_j q_j^2 t_{0,j}}{m_j \left(1 + \omega^2 t_{0,j}^2\right)} - \mathrm{i} \frac{n_j q_j^2 \omega t_{0,j}^2}{m_j \left(1 + \omega^2 t_{0,j}^2\right)} \right) \mathbf{E} = \left(\sigma_{1,j} - \mathrm{i}\sigma_{2,j}\right) \mathbf{E}, \tag{2.6}$$

where the real part σ_1 denotes ohmic dissipation in the normal state, whereas the imaginary part σ_2 accounts for the superconducting charge carrier's inertia.

2.1.2 Fluxoid and magnetic flux quantization

Due to their macroscopic wave function, which has to be single-valued for integration along a closed path $\partial\mathbf{S}$ [Lon50], Fritz London predicted a "fluxoid" quantization in multiple connected superconductors. The wave function of Eq. (2.3) is invariant for phase changes in integer multiples of 2π, such that the closed line integral of the phase gradient holds

$$\oint_{\partial\mathbf{S}} \nabla\theta \mathrm{d}\mathbf{l} = \lim_{\mathbf{r}_2\to\mathbf{r}_1} \int_{\mathbf{r}_1}^{\mathbf{r}_2} \nabla\theta \mathrm{d}\mathbf{l} = \lim_{\mathbf{r}_2\to\mathbf{r}_1} \left[\theta(\mathbf{r}_2,t) - \theta(\mathbf{r}_1,t)\right] = 2\pi n, \quad n \in \mathbb{Z}. \tag{2.7}$$

Thus, the closed line integral of the London equation (2.2c) yields the fluxoid

$$\Phi' = \int_S \mathbf{B}\mathrm{d}\mathbf{S} + \mu_0\lambda_\mathrm{L}^2 \oint_{\partial S} \mathbf{j}\mathrm{d}\mathbf{l} = n \cdot \frac{h}{q_\mathrm{s}}, \tag{2.8}$$

which is Bohr-Sommerfeld-like quantized in multiples of the superconducting flux quantum $\Phi_0 = h/q_\mathrm{s}$. In case of bulk superconductors, well inside the superconductor, screening currents vanish according to the Meissner-Ochsenfeld effect and the fluxoid quantization of Eq. (2.8) results in a magnetic flux quantization $\int \mathbf{B}\mathrm{d}\mathbf{S} = \Phi = n\Phi_0$. This means that multiple connected superconductors can merely enclose integer multiples of the superconducting flux quantum Φ_0.

Fritz London assumed erroneously $q_\mathrm{s} = -e$, but experiments, e.g., by Robert Doll and Martin Näbauer [DN61] as well as Bascom Sine Deaver Jr. and William Martin Fairbank [DF61] in 1961, showed the superconducting charge quantum to be $q_\mathrm{s} \equiv -2e$. This supports the idea of Cooper pairs retrospectively, on which the microscopic BCS theory bases. The superconducting magnetic flux quantum's current value in the literature is [Tie+21]

$$\Phi_0 \equiv \frac{h}{2e} = 2.067\,833\,848 \times 10^{-15}\,\mathrm{V\,s}. \tag{2.9}$$

2.1.3 Ginzburg-Landau theory

The previously mentioned observation, that the Meissner-Ochsenfeld effect is thermodynamically reversible, reveals that superconductivity is a separate thermodynamic state. This motivated Witali Lasarewitsch Ginzburg and Lew Dawidowitsch Landau to describe the superconductor's second order phase transition by Landau's theory of phase transitions [Lan37], extended by the superconductor's response to external magnetic fields [LG50; GL09]. The Landau theory relates the system's changing properties with its breaking symmetry that is characterized by the occurrence of a complex, spatially varying order parameter $\psi(\mathbf{r})$. As per definition of second order phase transitions, this order parameter exists exclusively in the less symmetric phase, approaches zero continuously at the phase transition and its square turns out to be the superconducting charge carrier density n_s. As the order parameter is small near the phase transition, the free energy without external magnetic field can be expanded into a symmetric power series

$$\mathcal{F}_s = \int f_s \mathrm{d}V = \int f_n + \alpha_{GL}|\psi|^2 + \frac{\beta_{GL}}{2}|\psi|^4 + \mathcal{O}\left(|\psi|^6\right)\mathrm{d}V. \tag{2.10}$$

Here, f_n is the free energy density of the normal conducting state that is assumed to be constant. The occurrence of superconductivity as the energetically favorable state below T_c necessitates $\alpha_{GL} < 0$ and $\beta_{GL} > 0$ in order to fulfill the energy lowering $f_s < f_n$ at a finite nonzero order parameter (see Fig. 2.1 a)). Then, the equilibrium states at the local minima, which hold $\frac{\partial f_s}{\partial|\psi|}\Big|_{\psi=\psi_\infty} = 0$, yield $\beta_{GL} = -\frac{\alpha_{GL}}{|\psi_\infty|^2}$ with the order parameter of the spatially homogeneous equilibrium ψ_∞.

As superconductors expel external magnetic fields from their interior, their Gibbs free energy density g_s is increased by the magnetic work $-\int \mathbf{M}\mathrm{d}\mathbf{B}_a$ in the presence of external magnetic fields. In case of perfect diamagnetism ($\chi_m = -1$), the magnetization is $\mathbf{M} = -\mathbf{H} = -\mathbf{B}_a/\mu_0$ and the field expulsion work becomes $\mathbf{B}_a^2/2\mu_0$ [GC34]. The induced currents are further associated with the charge carriers' kinetic energy $\frac{\hat{p}^2}{2m_s}$, where $\hat{p}$ is the canonical momentum $-\mathrm{i}\hbar\nabla - q_s\mathbf{A}$ derived from the principle of minimal coupling. All in all, for superconductors the Gibbs free energy density functional reads

$$g_s[\psi, \mathbf{A}] = g_n + \alpha_{GL}|\psi|^2 + \frac{\beta_{GL}}{2}|\psi|^4 + \frac{1}{2m_s}|(-\mathrm{i}\hbar\nabla - q_s\mathbf{A})\,\psi|^2 + \frac{\mathbf{B}_a^2}{2\mu_0}. \tag{2.11}$$

The order parameter in the equilibrium is obtained by minimizing the Gibbs free energy both with respect to variations in the order parameter $\delta\psi$ (while ψ^* is regarded independently) and the vector potential $\delta\mathbf{A}$ and results in the two Ginzburg-Landau equations

$$0 = \alpha_{GL}\psi + \beta_{GL}|\psi|^2\psi + \frac{1}{2m_s}\left(-\mathrm{i}\hbar\nabla - q_s\mathbf{A}\right)^2\psi, \tag{2.12a}$$

$$\begin{aligned}\mathbf{j}_s &= \frac{q_s\hbar}{2\mathrm{i}m_s}\left(\psi^*\nabla\psi - \psi\nabla\psi^*\right) - \frac{q_s^2}{m_s}\psi^*\psi\mathbf{A}\\ &= \frac{q_s}{m_s}\mathrm{Re}(\psi^*\left(-\mathrm{i}\hbar\nabla - q_s\mathbf{A}\right)\psi).\end{aligned} \tag{2.12b}$$

For homogeneous superconductors, the whole ensemble of superconducting electrons can be treated as macroscopic quantum state with one single wave function

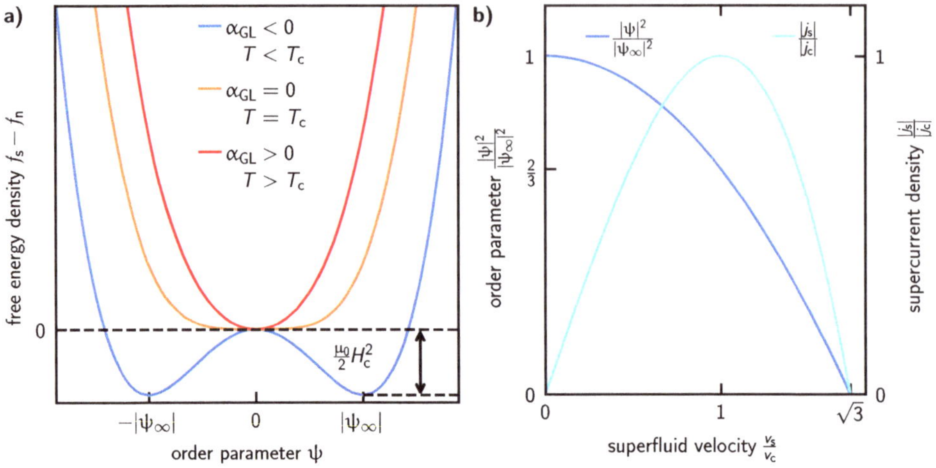

Figure 2.1: Ginzburg-Landau free energy and field dependence of the emerging order parameter. **a)** Symmetry breaking at the normal to superconducting phase transition. For $\alpha_{GL} > 0$ the equilibrium state is characterized by the absence of the order parameter $\psi = 0$, which is the case for $T > T_c$, while the occurrence of a finite order parameter $\psi = \psi_\infty$ below $T < T_c$ requires $\alpha_{GL} < 0$. The resulting energy lowering (difference of the super- and normal conducting free energy densities $f_s - f_n$) can for instance be suppressed by applied magnetic fields, and thus overcome by exceeding the critical value $H_c = \sqrt{\frac{\alpha_{GL}{}^2}{\mu_0 \beta_{GL}}}$. **b)** The order parameter ψ is suppressed in strong electric fields, which results in a nonlinear supercurrent density j_s with a maximal (critical) value j_c.

$\psi(\mathbf{r}) = |\psi| e^{i\theta(\mathbf{r})}$ according to Fritz London's postulate [Lon48][2] and the supercurrent density of Eq. (2.12b) results in

$$\mathbf{j}_s = \frac{q_s}{m_s} |\psi|^2 \left(\hbar \nabla \theta - q_s \mathbf{A}\right) = q_s |\psi|^2 \mathbf{v}_s. \tag{2.13}$$

The comparison with the London equation (2.2c) identifies the order parameter to be given by the superconducting charge carrier density $|\psi|^2 = n_s$. For weak perturbation fields, the last two terms of g_s in Eq. (2.11) are negligibly small, so that the thermodynamic equilibrium state emerges in the potential minima at $|\psi| = |\psi_\infty| = \sqrt{-\alpha_{GL}/\beta_{GL}}$, as illustrated in Fig. 2.1 a). In case of strong fields, however, the superfluid's kinetic energy $\frac{1}{2} n_s m_s |\mathbf{v}_s|^2$ can no longer be neglected. Since the kinetic properties of the superconducting charge carriers will be of particular interest for the experiments in this work, they will be considered in more detail below. Optimizing the homogeneous order parameter of the Ginzburg-Landau Gibbs free energy

$$g_s[\psi, \mathbf{A}] = g_n + \alpha_{GL} |\psi|^2 + \frac{\beta_{GL}}{2} |\psi|^4 + \frac{1}{2} m_s |\mathbf{v}_s|^2 |\psi|^2 + \frac{\mathbf{B}_a^2}{2\mu_0} \tag{2.14}$$

[2] This assumption was justified by means of microscopic BCS theory by Lev Petrovich Gor'kov [Gor59].

yields $|\psi|^2 = |\psi_\infty|^2 \left(1 - \frac{m_s|\mathbf{v}_s|^2}{|\alpha_{GL}|}\right)$ [Tin04] and consequently a nonlinear supercurrent density $\mathbf{j}_s = q_s|\psi|^2\mathbf{v}_s$ by inserting in Eq. (2.13). Figure 2.1 b) shows both the reduced charge carrier densities with increased superfluid velocities v_s. Depairing thermodynamically is thus favored, and the maximum supercurrent density is reached for $|\alpha_{GL}| = \frac{3m_s v_s^2}{2}$. This means that there is a maximal (critical) current density

$$\mathbf{j}_c = \frac{2}{3} q_s |\psi_\infty|^2 \mathbf{v}_c \tag{2.15}$$

above which superconductivity collapses. With the corresponding critical velocity v_c, the supercurrent density results in

$$\mathbf{j}_s = |\psi_\infty|^2 q_s \mathbf{v}_s \left(1 - \frac{|\mathbf{v}_s|^2}{3v_c^2}\right). \tag{2.16}$$

As mentioned above, the second Ginzburg-Landau equation (2.12b) corresponds to the London equation (2.2c) with the characteristic London penetration depth in the case of a homogeneous superconducting charge carrier density, and thus order parameter. Furthermore, the Ginzburg-Landau theory exhibits a second characteristic length scale of superconductors, on which $\psi(\mathbf{r})$ may vary spatially.

In the absence of magnetic fields and for small order parameters, e.g., near T_c, the $\mathcal{O}\left(|\psi|^2\psi\right)$ term is negligible and the first Ginzburg-Landau equation (2.12a) becomes $\nabla^2\psi = \frac{2m_s|\alpha_{GL}|}{\hbar^2}\psi$. This means that small perturbations $\delta\psi$ decay exponentially on the characteristic length scale, the so-called Ginzburg-Landau coherence length

$$\xi_{GL} = \sqrt{\frac{\hbar^2}{2m_s|\alpha_{GL}|}}. \tag{2.17}$$

These two length scales, the coherence length ξ_{GL} and the London penetration depth λ_L, are main results of the Ginzburg-Landau theory and important quantities to describe the different properties of superconductors. For instance, the ratio λ_L/ξ_{GL} defines whether superconductivity collapses abruptly above one critical magnetic field H_c (type-I superconductor with $\lambda_L/\xi_{GL} \leq 1/\sqrt{2}$) or the superconductor allows magnetic flux to penetrate partially above the first critical field H_{c1} and becomes completely normal conducting not until the second critical field $H_{c2} > H_{c1}$ (type-II superconductor with $\lambda_L/\xi_{GL} > 1/\sqrt{2}$). In the so-called Abrikosov or Shubnikov phase between H_{c1} and H_{c2} of the latter case, flux quanta with normal conducting cores get in the superconductor, and thus make for imperfect diamagnetism ($-1 < \chi_m < 0$) [DP19]. This second critical field can exceed the first critical field by orders of magnitudes [MW18].

The phenomenological Ginzburg-Landau theory covers the superconductor's unique behavior, namely electrodynamics and thermodynamics, using a macroscopic wave function of conduction electrons. However, it provides neither a microscopic explanation of their origin nor of the occurrence of superconductivity in general.

2.1.4 Microscopic Bardeen-Cooper-Schrieffer theory

The microscopic theory of superconductivity by John Bardeen, Leon Neil Cooper and John Robert Schrieffer (BCS) heralded a breakthrough in basic understanding of superconductivity. This theory is founded on a boson-like ground state of condensed, paired conduction electrons originating from an attractive electron-electron interaction, as it would not be possible for single electrons according to the Pauli exclusion principle [Tin04; Sch19; DP19]. Despite the repulsive Coulomb interaction between two isolated electrons, a virtual phonon can mediate a net attractive interaction in the solid body of conventional superconductors.[3] This counterintuitive exchange mechanism is inspired by evidencing the isotope effect of superconductors [Max50; Rey+50; Frö50; Bar50], whose transition temperature depends on the isotope mass as $T_c \propto 1/\sqrt{m_u}$ and therefore on the phonon frequency. Since phonon energies are limited to the Debye frequency $\hbar\omega_D$, the interaction happens between the Fermi energy E_F and $E_F + \hbar\omega_D$ and is most effective for electrons with opposite momentum and spin $|\mathbf{k}\downarrow, -\mathbf{k}\uparrow\rangle$. [4] The energy of these two electrons, a so-called Cooper pair, is lowered by approximately [Tin04]

$$\Delta E \approx -2\hbar\omega_D e^{-\frac{2}{\mathcal{N}_n(E_F)V}} \tag{2.18}$$

compared to two "free" electrons of energy $2E_F$. The Hamiltonian of the whole ensemble of superconducting electrons in the volume V holds

$$\mathcal{H}_{BCS} = \sum_{\mathbf{k},\sigma} \varepsilon_{\mathbf{k}} \hat{c}^\dagger_{\mathbf{k},\sigma} \hat{c}_{\mathbf{k},\sigma} + \sum_{\mathbf{k},\mathbf{k}'} \frac{g_{\mathbf{k},\mathbf{k}'}}{V} \hat{c}^\dagger_{\mathbf{k}',\uparrow} \hat{c}^\dagger_{-\mathbf{k}',\downarrow} \hat{c}_{-\mathbf{k},\downarrow} \hat{c}_{\mathbf{k},\uparrow}, \tag{2.19}$$

where $\hat{c}^\dagger_{\mathbf{k},\sigma}$ and $\hat{c}_{\mathbf{k},\sigma}$ denote electron's creation and annihilation operators, respectively.[5] The first term represents the electron's kinetic energy $\varepsilon_{\mathbf{k}} = \frac{\hbar^2\mathbf{k}^2}{2m} - \mu$ with

---

[3] It should be noted that this net attractive interaction does not necessarily need to be mediated by a virtual phonon. This is only the case for "conventional" superconductors, but for instance $T_c \gtrsim 90\,\mathrm{K}$ of high-temperature superconductors cannot be explained by a phononic interaction [Czy17].

[4] These singlet s-wave Cooper pairs with zero total spin and orbital angular momentum $S = L = 0$ are the simplest case, but for instance singlet d-wave Cooper pairs characterize high T_c superconductors [SB07; BK03; BK04].

[5] These fermionic creation and annihilation operators fulfill the typical commutation rules $\left\{\hat{c}_{\mathbf{k},\sigma}, \hat{c}^\dagger_{\mathbf{k}',\sigma'}\right\} = \delta_{\mathbf{k},\mathbf{k}'}\delta_{\sigma,\sigma'}$ and $\left\{\hat{c}^\dagger_{\mathbf{k},\sigma}, \hat{c}^\dagger_{\mathbf{k}',\sigma'}\right\} = \left\{\hat{c}_{\mathbf{k},\sigma}, \hat{c}_{\mathbf{k}',\sigma'}\right\} = 0$.

the chemical potential μ. The Cooper pairing mechanism in the second term is described by the attractive interaction $g_{\mathbf{k},\mathbf{k}'}$ of two electrons that are scattered from momentum $\pm\mathbf{k}$ to $\pm\mathbf{k}'$. These Cooper pairs typically extend over a range of 100 nm–1000 nm, overlapping with $\sim 10^9$ Cooper pairs. Therefore, the interaction among Cooper pairs can be considered as interaction with a Cooper pair mean-field $\Delta_{\mathbf{k}} = -\frac{1}{V}\sum_{\mathbf{k}'} g_{\mathbf{k},\mathbf{k}'} \langle \hat{c}_{-\mathbf{k}',\downarrow}\hat{c}_{\mathbf{k}',\uparrow}\rangle$ [Tin04; Czy17].

Moreover, the many-particle wave function describing the superconducting system simplifies in the mean-field approximation to [Tin04; Sch19]

$$|\Psi_{\mathrm{BCS}}\rangle = \prod_{\mathbf{k}} \left(u_{\mathbf{k}} + v_{\mathbf{k}}\hat{c}^{\dagger}_{\mathbf{k},\uparrow}\hat{c}^{\dagger}_{-\mathbf{k},\downarrow}\right)|\Psi_0\rangle = \prod_{\mathbf{k}} \left(|u_{\mathbf{k}}| + |v_{\mathbf{k}}|\mathrm{e}^{\mathrm{i}\theta}\hat{c}^{\dagger}_{\mathbf{k},\uparrow}\hat{c}^{\dagger}_{-\mathbf{k},\downarrow}\right)|\Psi_0\rangle\,. \tag{2.20}$$

Here, $|v_{\mathbf{k}}|^2$ denotes the probability of the singlet pair $|\mathbf{k}\downarrow, -\mathbf{k}\uparrow\rangle$ being occupied, $|u_{\mathbf{k}}|^2 = 1 - |v_{\mathbf{k}}|^2$ the probability being unoccupied, respectively. The coefficients $u_{\mathbf{k}}, v_{\mathbf{k}} \in \mathbb{C}$ can be chosen as $u_{\mathbf{k}} = |u_{\mathbf{k}}|$ and $v_{\mathbf{k}} = |v_{\mathbf{k}}|\mathrm{e}^{\mathrm{i}\theta}$, since the creation operator $\hat{c}^{\dagger}_{\mathbf{k},\uparrow}\hat{c}^{\dagger}_{-\mathbf{k},\downarrow}$ creates one Cooper pair with a certain phase $\mathrm{e}^{\mathrm{i}\theta}$ from the vacuum state $|\Psi_0\rangle$. This reveals that the BCS ground state is a coherent many-particle state in which all Cooper pairs are in a single common quantum state. Due to the long-range coherence of the BCS ground state, superconductors can effectively be described by the macroscopic wave function of Eq. (2.3), as postulated by London [Lon48] and assumed in the Ginzburg-Landau theory (see Sec. 2.1.3).

As can be seen in the BCS-Hamiltonian of Eq. (2.19), electrons are created and annihilate in pairs exclusively, so that their creation and annihilation operators need to be decoupled from each other by means of a linear transformation, the so-called Bogolyubov-Valatin transformation. The new fermionic operators [BTŠ58; Val58] are given by

$$\begin{pmatrix}\hat{\gamma}_{\mathbf{k},\uparrow}\\ \hat{\gamma}^{\dagger}_{-\mathbf{k},\downarrow}\end{pmatrix} = \begin{pmatrix} u_{\mathbf{k}} & -v_{\mathbf{k}}\\ v^{*}_{\mathbf{k}} & u^{*}_{\mathbf{k}}\end{pmatrix}\begin{pmatrix}\hat{c}_{\mathbf{k},\uparrow}\\ \hat{c}^{\dagger}_{-\mathbf{k},\downarrow}\end{pmatrix} \Leftrightarrow \begin{pmatrix}\hat{c}_{\mathbf{k},\uparrow}\\ \hat{c}^{\dagger}_{-\mathbf{k},\downarrow}\end{pmatrix} = \begin{pmatrix} u^{*}_{\mathbf{k}} & v_{\mathbf{k}}\\ -v^{*}_{\mathbf{k}} & u_{\mathbf{k}}\end{pmatrix}\begin{pmatrix}\hat{\gamma}_{\mathbf{k},\uparrow}\\ \hat{\gamma}^{\dagger}_{-\mathbf{k},\downarrow}\end{pmatrix}. \tag{2.21}$$

In case of phonon mediated singlet s-wave Cooper pairs, the attractive electron interaction $g_{\mathbf{k},\mathbf{k}'}$ is bound to the Debye energy around the Fermi surface and is further isotropic [Col15], such that

$$g_{\mathbf{k},\mathbf{k}'} = \begin{cases} -g, & |\varepsilon_{\mathbf{k}}|, |\varepsilon_{\mathbf{k}'}| \leq \hbar\omega_{\mathrm{D}} \\ 0, & \text{otherwise}\end{cases} \qquad \Delta_{\mathbf{k}} = \begin{cases} \Delta, & |\varepsilon_{\mathbf{k}}| \leq \hbar\omega_{\mathrm{D}} \\ 0, & \text{otherwise}\end{cases}. \tag{2.22}$$

The diagonalized effective BCS-Hamiltonian finally holds

$$\mathcal{H}^{\mathrm{eff}}_{\mathrm{BCS}} = \sum_{\mathbf{k},\sigma} E_{\mathbf{k}}\hat{\gamma}^{\dagger}_{\mathbf{k},\sigma}\hat{\gamma}_{\mathbf{k},\sigma} + \sum_{\mathbf{k}}(\varepsilon_{\mathbf{k}} - E_{\mathbf{k}}) + \frac{V\Delta^2}{g}. \tag{2.23}$$

The second and third term in Eq. (2.23) denote the constant energy gained due to the creation of Cooper pairs (condensation energy). The introduced energy

$$E_{\mathbf{k}} = \sqrt{\varepsilon_{\mathbf{k}}^2 + |\Delta|^2} \tag{2.24}$$

emphasizes that Δ manifests as a gap in the superconductor's energy spectrum even at the Fermi level, where $\varepsilon_{\mathbf{k}} = 0$. This energy gap and the macroscopic wave function are the reason for the superconductor's vanishing DC resistance up to the critical current, since the absence of free states does not allow for scattering with an equivalent energy transfer less than 2Δ. The size of the superconducting energy gap varies among different materials and at absolute zero temperature is given by

$$\Delta_0 = \frac{\hbar\omega_{\mathrm{D}}}{\sinh\left(\frac{1}{\mathcal{N}(E_{\mathrm{F}})g}\right)} \approx 2\hbar\omega_{\mathrm{D}} \mathrm{e}^{-\frac{1}{\mathcal{N}_{\mathrm{n}}(E_{\mathrm{F}})g}} = \pi \mathrm{e}^{-\gamma} k_{\mathrm{B}} T_{\mathrm{c}} \approx 1.764 k_{\mathrm{B}} T_{\mathrm{c}} \tag{2.25}$$

with Euler's constant $\gamma \approx 0.5772$. Equation (2.25) explains the superconducting isotope effect since the superconductor's gap energy is both directly proportional to its critical temperature T_{c} and the Debye frequency.

Excitations that exceed the minimum energy of $2|\Delta|$ break a Cooper pair into two Bogoliubov quasiparticles, as described in the first term of Eq. (2.23) by the operators $\hat{\gamma}_{\mathbf{k},\sigma}$.[6] The quasiparticles' density of states, as visualized in Fig. 2.2 a), results in [Tin04; Kit05]

$$\mathcal{N}_{\mathrm{qp}}(E_{\mathbf{k}}) = \frac{V}{(2\pi)^3} \int\limits_{E=\mathrm{const}} \frac{\mathrm{d}S_{\mathbf{k}}}{|\boldsymbol{\nabla}_{\mathbf{k}} E_{\mathbf{k}}|} = \mathcal{N}_{\mathrm{n}}(E_{\mathrm{F}}) \frac{E_{\mathbf{k}}}{\sqrt{E_{\mathbf{k}}^2 - \Delta^2}} \Theta(E_{\mathbf{k}} - \Delta) \tag{2.26}$$

with the normal conducting density of states at the Fermi level $\mathcal{N}_{\mathrm{n}}(E_{\mathrm{F}}) = \frac{m|\mathbf{k}_{\mathrm{F}}|}{2\pi^2\hbar^2}$. From the quasiparticles' density of states, a band diagram can be deduced that features no contribution of quasiparticles to the conductance at zero temperature and a thermally populated conductance band above the gap for finite temperatures.[7]

6 It is conspicuous that the operators couple electrons and holes, which is why these new fermions are called quasiparticles.

7 Cooper pairs at the Fermi energy E_{F} might be confusing since their energy is lowered by the superconducting gap according to Eq. (2.18). In the picture of band diagrams that illustrate fermionic densities of states and possible gaps, the boson-like Cooper pairs that condensate in a common ground state can not be included in the same manner. To illustrate transport properties, however, a combined band diagram of Cooper pairs and their quasiparticle excitations is useful. First, the Fermi energy in superconductors at zero temperature is above a fully occupied band and below an empty band because there is no fermionic conductance band. Small changes within this gap do not change the superconducting behavior, so they can be considered fermionic insulators. However, they show perfect conductivity but carried by boson-like Cooper pairs. Second, the energy 2Δ is required to excite two quasiparticles, which were not in any other fermion states of the density of states before. Thus, Cooper pairs are below the lower edge of the upper unoccupied valence band by Δ.

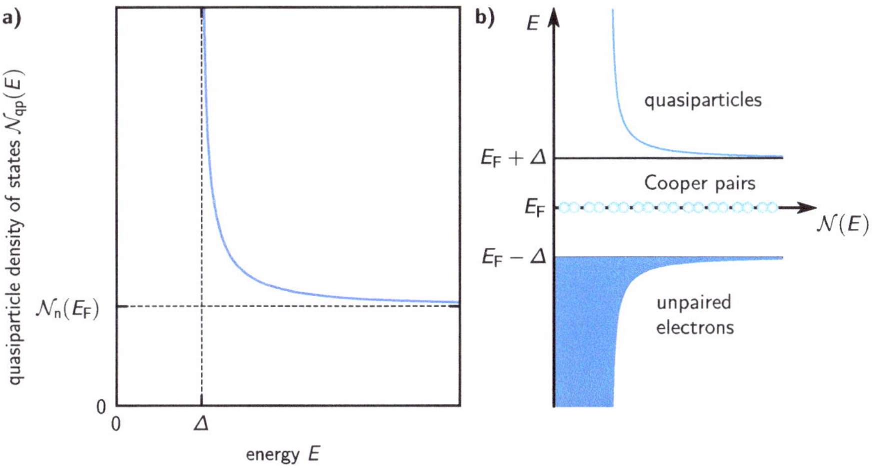

Figure 2.2: Quasiparticle density of states and the deduced superconductor's band structure. a) The quasiparticle density of states shows a superconducting gap of size 2Δ around the Fermi level, where there are no states available. However, near above the gap, far more states are possible compared to the normal conducting case, whose value $\mathcal{N}_n(E_F)$ is approached far above the gap. **b)** Against this backdrop, a band diagram can be designed for superconductors, which is a useful concept to describe the transport behavior of coupled superconductors such as the Josephson effect [NSS60; KW19].

2.2 Granular high-kinetic inductance superconductors

For the design of superconducting circuits, the basic components, namely resistors, capacitors, and inductors, need to be known. As superconductors stand out because of their zero DC resistance, resistors rather appear from normal conducting quasiparticles. Capacitors are dominated by their dielectric, and thus are not influenced by superconducting electrodes. However, the inductance of superconducting devices may feature additional terms compared to common normal conductors, which are addressed in this section.

2.2.1 Kinetic inductance

Like any other current-carrying conductor, superconducting strip lines exhibit a linear geometric inductance L_g, which depends on their actual shape and dimensions and stores energy in the surrounding B-field $\frac{1}{2\mu_0}\int \mathbf{B}^2 \mathrm{d}V = \frac{1}{2}L_g I^2$. This magnetic energy, however, is not the entire energy that is carried by the electric current since moving charge carriers themselves naturally hold kinetic energy $\frac{1}{2}\int n_s m_s |\mathbf{v}_s|^2 \mathrm{d}V$.

As current I and charge carrier velocity are proportionally to the first order, the kinetic part contributes to the total inductance, too. In normal conductors, the magnetic due exceeds the kinetic due by orders of magnitudes, except for very high frequencies $f \gtrsim 10^{13}\,\mathrm{Hz}$ [SMU97]. In thin superconducting strips or films, however, charge carriers can gain kinetic energies comparable or exceeding the magnetic energy even for $f \to 0$ because of the substantially higher charge carrier mobility [Ann10].

The fact that the energy, associated with a kinetic inductance, is spatially confined to the superconductor's interior leads to a material specific constant inductivity and a constant inductance per unit square for thin homogeneous films of fixed thickness d. In the latter case, the element's total inductance kinetic inductance $L_\mathrm{k}^\square = L_\mathrm{k}/N^\square$ scales with its number of squares $N^\square = l/w$, where l and w denote its length and width, respectively. In contrast, geometric inductances couple to the surrounding magnetic fields and depend on the element's actual shape and dimensions but not on its material. This is why the geometric inductance is usually referred to the constant inductance in terms per unit length $L'_\mathrm{g} = L_\mathrm{g}/l$.

By definition, the inductance relates the change in current through and the voltage drop across an element as $\mathbf{E}l = L_\mathrm{k}\frac{\mathrm{d}\mathbf{I}_\mathrm{s}}{\mathrm{d}t} = L_\mathrm{k}^\square \frac{l}{w} wd \frac{\mathrm{d}\mathbf{j}_\mathrm{s}}{\mathrm{d}t}$. The electric field $\mathbf{E}$, in turn, accelerates the superfluid $q_\mathrm{s}\mathbf{E} = \frac{\mathrm{d}\mathbf{p}_\mathrm{s}}{\mathrm{d}t} = m_\mathrm{s}\frac{\mathrm{d}\mathbf{v}_\mathrm{s}}{\mathrm{d}t}$ to increases its momentum $\mathbf{p}_\mathrm{s} = m_\mathrm{s}\mathbf{v}_\mathrm{s}$. This leads to the kinetic square inductance of a superconductor [MT69; CHS71; ASB89]

$$L_\mathrm{k}^\square = \frac{|\mathbf{E}|}{d}\left(\frac{\mathrm{d}|\mathbf{j}_\mathrm{s}|}{\mathrm{d}t}\right)^{-1} = \frac{m_\mathrm{s}}{q_\mathrm{s}d}\frac{\mathrm{d}|\mathbf{v}_\mathrm{s}|}{\mathrm{d}t}\left(\frac{\mathrm{d}|\mathbf{j}_\mathrm{s}|}{\mathrm{d}t}\right)^{-1} = \frac{m_\mathrm{s}}{q_\mathrm{s}d}\left(\frac{\mathrm{d}j_\mathrm{s}}{\mathrm{d}v_\mathrm{s}}\right)^{-1}. \tag{2.27}$$

As already discussed as part of the Ginzburg-Landau theory in Sec. 2.1.3, the homogeneous supercurrent density is described by the London theory $\mathbf{j}_\mathrm{s} = n_\mathrm{s}q_\mathrm{s}\mathbf{v}_\mathrm{s}$ with $n_\mathrm{s} = |\psi|^2 = |\psi_\infty|^2$ in the simplest case of thin superconducting films in weak perturbating fields [Tin04] and therefore the kinetic square inductance results in

$$L_\mathrm{k}^\square(0) = \frac{m_\mathrm{s}}{n_\mathrm{s}q_\mathrm{s}^2}\frac{1}{d} = \mu_0\frac{\lambda_\mathrm{L}^2}{d}. \tag{2.28}$$

In case of strong perturbations, however, the $\mathbf{j}_\mathrm{s}(\mathbf{v}_\mathrm{s})$ dependence becomes nonlinear, as given in Eq. (2.16) and calculating the kinetic square inductance of Eq. (2.27) results in [ASB89]

$$L_\mathrm{k}^\square(\mathbf{v}_\mathrm{s}) = \frac{m_\mathrm{s}}{|\psi_\infty|^2 q_\mathrm{s}^2 d\left(1-\frac{v_\mathrm{s}^2}{v_\mathrm{c}^2}\right)} = L_\mathrm{k}^\square(0)\left(1-\frac{\mathbf{v}_\mathrm{s}^2}{\mathbf{v}_\mathrm{c}^2}\right)^{-1}. \tag{2.29}$$

To use the more measurable quantities, supercurrent and critical current density, $\mathbf{j}_\mathrm{s}(\mathbf{v}_\mathrm{s})$ needs to be inverted along its stable branch. The Eqs. (2.15) and (2.16) relate

Table 2.1: Properties of several high-kinetic inductance superconductors. The kinetic sheet inductance $L_k^\square$ is both affected by critical temperature T_c as a measure of the superconducting gap energy and the normal state resistivity ρ_n according to Eq. (2.32). These quantities are reported from several circuits made from different high-kinetic inductance superconductors with comparable thicknesses.

material	ρ_n (μΩ cm)	T_c (K)	$L_k^\square$ [8] (pH sq^{-1})	references
NbN	~1000	~7.2	~80	[NBB19; Gra+08; Luo+14]
TiN	~450	~3.0	~235	[Led+10; She+18; Swe+13; Vis+10]
NbTiN	~200	~9.3	~35	[Bar+10; Sam+16; Haz+19]
MoGe	~220	~6.0	~45	[Dau+21; Bol+08]
AlO_x	~6000	~1.6	~2000	[Rot+16]

$\frac{\mathbf{j}_s}{\mathbf{j}_c} = \frac{3\mathbf{v}_s}{2\mathbf{v}_c} - \frac{\mathbf{v}_s^3}{2\mathbf{v}_c^3}$ and for small bias currents, where only the first order term is considered, the kinetic square inductance is approximated to [ASB89]

$$L_k^\square(j_s) \approx \frac{L_k^\square(0)}{1-\left(\frac{2j_s}{3j_c}\right)^2} \approx L_k^\square(0)\left(1+\left(\frac{2j_s}{3j_c}\right)^2+\left(\frac{2j_s}{3j_c}\right)^4+\mathcal{O}\left(j_s^6\right)\right). \tag{2.30}$$

Apart from this bias dependence, $L_k^\square(0) \propto n_s^{-1}$ strongly depends on the temperature since the superfluid density vanishes at the phase transition at $T \to T_c$. The exact temperature dependence is individual for different superconductors [Tin04], but can be approximated by the Gorter-Casimir two-fluid model $\frac{n_s(T)}{n_s(0)} = 1-\left(\frac{T}{T_c}\right)^4$. Finally, the kinetic square inductance is given by

$$L_k^\square(T, j_s) = L_k^\square(0)\frac{1}{1-\left(\frac{T}{T_c}\right)^4}\left(1+\left(\frac{2j_s}{3j_c}\right)^2+\left(\frac{2j_s}{3j_c}\right)^4+\mathcal{O}\left(j_s^6\right)\right). \tag{2.31}$$

In any superconductor, the kinetic inductance diverges close to T_c, where order parameter approaches zero, and plays an important role for any superconductor. At low temperatures far below T_c, kinetic inductance is only relevant in superconductors with small superfluid densities. This can be explained descriptively as follows: By driving a given bias current, less participating superconducting charge carriers are accelerated to antiproportionally faster velocities and their kinetic energy thus increases quadratically, such that the kinetic energy of the whole superfluid increases antiproportionally to n_s.

[8] at a film thickness of $d = 20\,\mathrm{nm}$

2.2.2 Mattis-Bardeen theory of granular superconductors

Due to their suppressed superfluid density, granular superconductor devices are propitious to achieve large values of kinetic inductance. In order to describe the kinetic inductance of circuits made from granular superconductors by means of their normal conducting properties, it is worth to look at kinetic inductance from another perspective of AC circuit analysis. The superconductor's AC response exhibits a negative susceptance $\mathrm{Im}(Z^{-1}) \propto \sigma_2$ (see Sec. 2.1.1), which corresponds to an inductance that accounts for the superfluid's inertia. Hence, the superconductor's complex admittance $\sigma_\mathrm{s} = \sigma_1 - \mathrm{i}\sigma_2$ defines its impedance by $\sigma_\mathrm{s}\frac{wd}{l} = \frac{1}{Z} = \frac{1}{R+\mathrm{i}\omega L_\mathrm{k}}$. Considering the superconductor's vanishing DC resistance, its impedance is exclusively given by its kinetic inductance $|Z| = \omega L_\mathrm{k}$, such that this kinetic inductance results in $L_\mathrm{k}^\square = \frac{1}{\omega\,\mathrm{Im}(\sigma_\mathrm{s})d}$.

In the local ($l_0 \ll \lambda_\mathrm{L}$), dirty ($l_0 \ll \xi$) and low-frequency limit ($hf \ll k_\mathrm{B}T$) this imaginary superconductivity $\mathrm{Im}(\sigma_\mathrm{s})$ is related to the real normal state conductivity σ_n by the Mattis-Bardeen theory as $\frac{\mathrm{Im}(\sigma_\mathrm{s})}{\sigma_\mathrm{n}} = \frac{\pi\Delta(T)}{hf}\tanh\left(\frac{\Delta(T)}{2k_\mathrm{B}T}\right)$. [MB58; Tin04]. Thus, the kinetic sheet inductance

$$L_\mathrm{k}^\square = \frac{1}{\omega\,\mathrm{Im}(\sigma_\mathrm{s})d} = \frac{\hbar R_\mathrm{n}^\square}{\pi\Delta(T)}\coth\left(\frac{\Delta(T)}{2k_\mathrm{B}T}\right) \xrightarrow{T\to 0} \frac{\hbar R_\mathrm{n}^\square}{\pi\Delta(0)} \tag{2.32}$$

is directly linked to the normal sheet resistance $\sigma_\mathrm{n} = \rho_\mathrm{n}^{-1} = (R_\mathrm{n}dw/l)^{-1} = (R_\mathrm{n}^\square d)^{-1}$. To reach large kinetic inductance values, superconductor with small superconducting gaps ($\propto T_\mathrm{c}$) and large normal state resistances are preferred. An incomplete compilation of different popular granular superconductors as well as their kinetic inductance values is given in Tab. 2.1.

2.3 Superconducting microwave circuits

Nowadays, superconducting quantum circuits are commonly manipulated and read out in the microwave regime [Gu+17]. Unlike the DC measurement scheme, RF coupling can prevent galvanic contacts to the environment that would be a substantial decoherence channel. Thereby, microwave resonators are used as suitable detectors for coupled quantum systems since their resonant frequency shifts dispersively depending on the quantum state of the coupled system [Bla+04; Wal+04].

2.3.1 Scattering matrix of a microwave network

A common technique of electrical engineering to describe a microwave signal's interaction with any circuit, representing a n-port network, is a $n \times n$ complex scattering matrix, which relates the complex output and input voltage at port i and port j by [Poz11]

$$S_{ij} = \left.\frac{V_i^{\text{out}}}{V_j^{\text{in}}}\right|_{V_k^{\text{in}}=0, k\neq j} . \tag{2.33}$$

This means that S_{ii} denotes the reflection coefficient of the port i, whereas S_{ij} is the transmission coefficient from port j to i. Throughout this work, microwave spectroscopy is done at a two-port network, as illustrated in Fig. 2.3 a), given by[9]

$$\begin{pmatrix} V_1^{\text{out}} \\ V_2^{\text{out}} \end{pmatrix} = \begin{pmatrix} S_{11} & S_{12} \\ S_{21} & S_{22} \end{pmatrix} \begin{pmatrix} V_1^{\text{in}} \\ V_2^{\text{in}} \end{pmatrix} \quad \Leftrightarrow \quad \mathbf{V}^{\text{out}} = S\mathbf{V}^{\text{in}}. \tag{2.34}$$

In the present experiments, both ports are connected with a continuous transmission line, whose characteristic impedance is typically terminated to $Z_0 = 50\,\Omega$, and this transmission line is shunted to ground by a load with an arbitrary impedance Z_l (see Fig. 2.3). For this so-called notch-type configuration of a parallel scatterer, Kirchhoff's laws yield the scattering matrix [Poz11; Bra18]

$$S^{\text{notch}} = \begin{pmatrix} \left(1 + \frac{2Z_l}{Z_0}\right)^{-1} & \left(1 + \frac{Z_0}{2Z_l}\right)^{-1} \\ \left(1 + \frac{Z_0}{2Z_l}\right)^{-1} & \left(1 + \frac{2Z_l}{Z_0}\right)^{-1} \end{pmatrix} . \tag{2.35}$$

To give a vivid description, the abrupt impedance mismatch in the transmission line due to the external load yields a part of the incident signal being reflected and the other part being transmitted. Thereby, complete reflection is obtained for $Z_l \to 0$, complete transmission for $Z_l \to \infty$, whereas the signal is split in equal parts in case of $Z_l = Z_0/2$.

2.3.2 Superconducting microwave resonators

One of the simplest loads featuring quantized excitations are quantum harmonic oscillators in the form of RLC-circuits. In this work, such RLC-resonators are coupled to a microwave feedline on the one hand and to long Josephson junctions

[9] The matrices $S^{\mathsf{T}} = S$ of the circuits studied in this work are symmetric, which means that interchanging the ports 1 and 2 will not vary the system's behavior. The whole measurement setup where these samples are embedded, however, include active components (amplifiers) and ferrites (isolators) that break the time reversal symmetry so that the S-matrix is no longer symmetric.

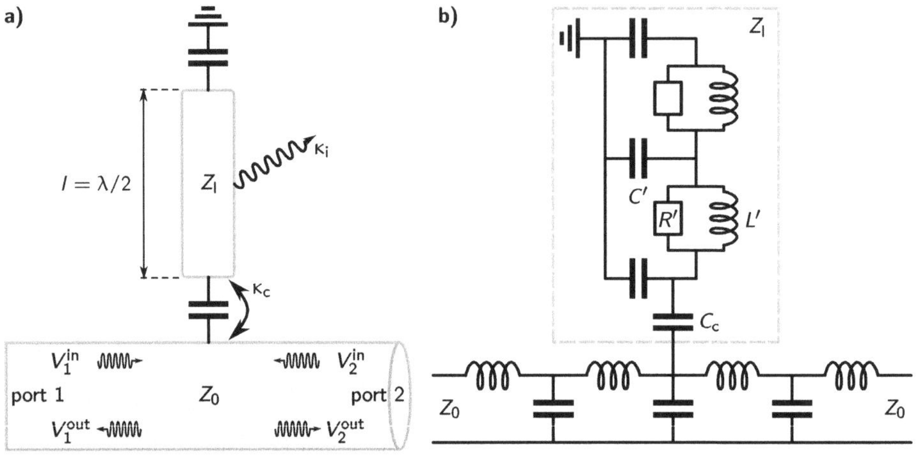

Figure 2.3: Schematic of a half-wavelength notch-type resonator embedded in a two port-network. **a)** The particular notch-type configuration consists of a *LC* transmission line with specific impedance $Z_0 = \sqrt{L/C}$ that is shunted to ground via a load with impedance Z_l. In the present case, this load is a terminated transmission line whose two open ends define a distributed $\lambda/2$ resonator. The two-port network of this assembly is characterized by relating the incident and scattered microwaves at both ports of the transmission line. The coupling between feedline and resonator is assumed to be capacitive with loss rate κ_c and the resonator itself has an internal linewidth κ_i. **b)** As a schematic circuit, the infinite transmission line is modeled as an inductive waveguide that is capacitively shunted to ground. The distributed *RLC*-resonator has a well-defined length, whose impedance mismatch at the open ends forms a resonating cavity.

on the other hand to investigate single photon excitations in these junctions. In order to probe these extended junctions, open-ended distributed $\lambda/2$ resonators are suitable since their current distribution is defined by the excited mode. In the center part of the resonator, $\lambda/2$ resonators exhibit current nodes and antinodes for odd and even harmonics, respectively, that probe the junctions differently. A $\lambda/2$ resonator itself is formed from a waveguide of a certain length l, which constrains resonating wavelengths to integer multiples of $\lambda/2$. Neglecting the frequency shift induced by the coupling capacitors, the resonant frequency of the nth harmonic of a $\lambda/2$-resonator holds

$$\omega_n^{\text{r}} = n \cdot \omega_1^{\text{r}} = n \cdot \frac{1}{2l\sqrt{L'C'}}. \tag{2.36}$$

The fundamental frequency ω_1^{r} depends on the inductance per unit length L' and capacitance per unit length C'. Since this quantity does not account for dissipation, quality factors are introduced, describing the ratio between the total energy stored in the resonating system E_{tot} and the energy dissipated per cycle $E_{\text{diss}} = P_{\text{loss}}/\omega^{\text{r}}$.

$$Q = \omega^{\text{r}} \frac{E_{\text{tot}}}{P_{\text{loss}}} = \frac{\omega^{\text{r}}}{\kappa}. \tag{2.37}$$

Here, κ denotes the inverse photon lifetime in a linear resonator, meaning that the excited resonator decays into its ground state at a rate $\kappa/2$. In the frequency domain, κ corresponds to the full width at half maximum of a resonator's Lorentzian shaped amplitude signal. For the whole load, there are two main loss channels: an internal due originating in ohmic dissipation and dielectric losses and a coupling due to other systems as e.g., the transmission line. For that reason, the total quality factor of the whole load Q_l composes of an internal quality factor Q_i and a coupling quality factor Q_c. Since decay rates κ add linearly, the loaded quality factor is given by

$$Q_\mathrm{l}^{-1} = Q_\mathrm{i}^{-1} + Q_\mathrm{c}^{-1} \qquad \kappa_\mathrm{l} = \kappa_\mathrm{i} + \kappa_\mathrm{c}. \tag{2.38}$$

In the notch-type configuration, the load's frequency-dependent input impedance can be expressed in terms of the internal and coupling quality factor as well as the frequency detuning from the resonant frequency [Bra18]

$$Z_\mathrm{in}^\mathrm{notch}(\omega) = Z_0 \left(\frac{Q_\mathrm{c}}{2Q_\mathrm{i}} + \mathrm{i}Q_\mathrm{c}\frac{\omega - \omega^\mathrm{r}}{\omega^\mathrm{r}} \right). \tag{2.39}$$

The S-parameters of ideal notch-type resonators can be found by inserting Eq. (2.39) into Eq. (2.35). Since the scattered signal propagates to both ports, $S_{11} + S_{21} = 1$ holds. In the following only S_{21} is treated since this is the later measured matrix entry and as it implies $S_{11} = 1 - S_{21}$. The complex signal is usually given in the polar form with amplitude and phase

$$S_{21}^\mathrm{notch} = \left(1 - \frac{Q_\mathrm{l}/Q_\mathrm{c}}{1 + 2\mathrm{i}Q_\mathrm{l}\frac{\omega - \omega^\mathrm{r}}{\omega^\mathrm{r}}} \right) = 1 - S_{11}, \tag{2.40a}$$

$$\left| S_{21}^\mathrm{notch} \right|^2 = 1 - \frac{1 - \left(Q_\mathrm{l} - Q_\mathrm{c}\right)^2 / Q_\mathrm{c}^2}{1 + 4Q_\mathrm{l}^2 \left(\frac{\omega - \omega^\mathrm{r}}{\omega^\mathrm{r}}\right)^2}, \tag{2.40b}$$

$$\arg\left(S_{21}^\mathrm{notch} \right) = -\arctan\left(2Q_l \frac{\omega - \omega_r}{\omega_r} \right) - \arctan\left(\frac{2Q_l Q_c}{Q_l - Q_c} \frac{\omega - \omega_r}{\omega_r} \right). \tag{2.40c}$$

Figure 2.4 shows both the complex plane, the amplitude, and the phase of S_{21} of a notch-type resonator. In the complex plane, the S_{21}-parameter manifests as a circle with diameter $Q_\mathrm{l}/Q_\mathrm{c}$ around the center, located at $1 - Q_\mathrm{l}/2Q_\mathrm{c}$. The Lorentzian shaped squared amplitudes $|S_{21}(\omega)|^2$ correspond to the Fourier transform of an exponential decay in the time domain and exhibit a full width at half maximum of κ_l. As can be seen in Fig. 2.4, the interplay of Q_i and Q_c plays a key role in the response of resonating systems that are usually categorized in three cases:

- under-coupled ($Q_\mathrm{i} < Q_\mathrm{c}$): Here, most of the energy is lost internally resulting in a weak signal both in amplitude and phase, and thus in a circle with a small diameter and a large distance from the origin

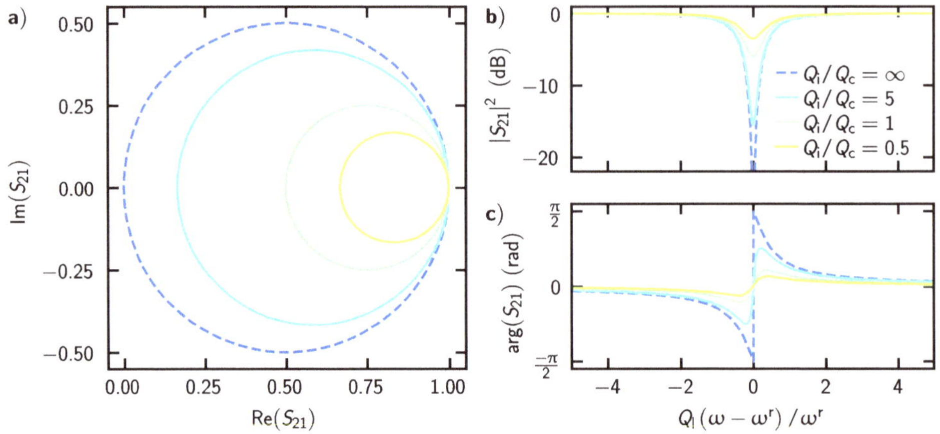

Figure 2.4: S_{21}-parameter of a notch-type resonator for different coupling regimes. In the complex plane **a)**, frequency sweeps around the resonant frequency feature circles, starting and ending at the off-resonant point $S_{21} = 1$. The the diameter of the circle increases with the ratio of $Q_\mathrm{i}/Q_\mathrm{c}$ and tends to 1 for $Q_\mathrm{i}/Q_\mathrm{c} \to \infty$ (dashed). The squared amplitude **b)** shows Lorentzian dips, the deeper the larger $Q_\mathrm{i}/Q_\mathrm{c}$ and the phase signal **c)** maximally rolls off from $-\pi/2$ to $\pi/2$ for $Q_\mathrm{i}/Q_\mathrm{c} \to \infty$.

- critically coupled ($Q_\mathrm{i} = Q_\mathrm{c}$): Here, half the energy is lost internally and the other half by coupling such that the amplitude on resonance equals $|S_{21}| = 0.5$ (-6 dB)
- over-coupled ($Q_\mathrm{i} > Q_\mathrm{c}$): Here, most of the energy is lost during coupling, resulting in the strongest signals with maximal diameter 1, zero transmission amplitude and a phase roll-off from $-\pi/2$ to $\pi/2$.

The complex-valued S_{21} data are experimentally acquired with the two-port vector network analyzer (VNA) and analyzed with a circle-fit routine (see Fig. 2.5), which provides values for the quality factors Q_l, Q_i, and Q_c [Pro+15]. For this purpose, Eq. (2.40) for an ideal notch-type resonator is adapted to the experimental setup. The environment is considered by an overall factor $a\mathrm{e}^{\mathrm{i}\alpha_0}\mathrm{e}^{-\mathrm{i}\omega\tau_\mathrm{el}}$ that includes attenuation and amplification of the whole signal a, a global phase offset α_0, and a cable delay τ_el denoting the signal's traveling time through the setup. Possible impedance mismatch is further taken into account by an asymmetry angle ϕ_Z that rotates the circle's center around the origin. This means that the resonance point is no longer on the real axis of the complex plane and the squared amplitude does not follow a clean Lorentzian. Having regard to these corrections, the transmission coefficient holds [Kha+12; Pro+15; Sch20]

$$S_{21}^{\mathrm{notch}} = a\mathrm{e}^{\mathrm{i}\alpha_0}\mathrm{e}^{-\mathrm{i}\omega\tau_\mathrm{el}}\left(1 - \frac{Q_\mathrm{l}/Q_\mathrm{c}\mathrm{e}^{\mathrm{i}\phi_Z}}{1 + 2\mathrm{i}Q_\mathrm{l}\frac{\omega-\omega^\mathrm{r}}{\omega^\mathrm{r}}}\right). \tag{2.41}$$

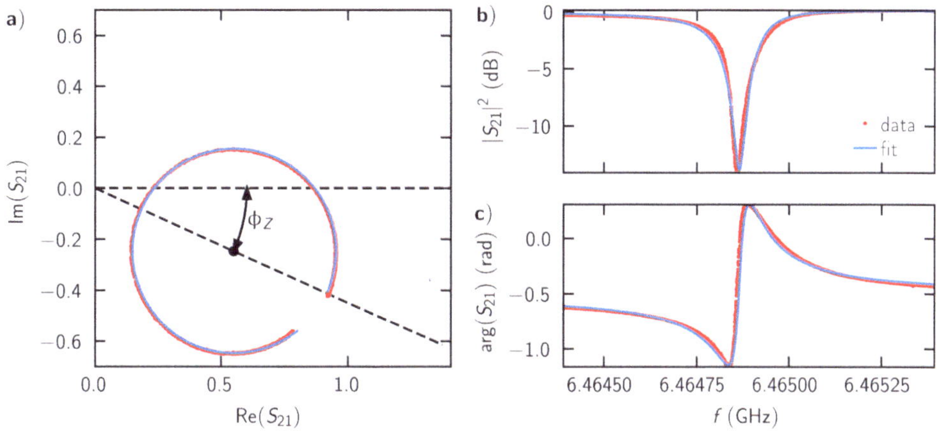

Figure 2.5: Measured and fitted S_{21}-parameter of a notch-type resonator. a) In the complex plane, the circle is rotated around the origin by the angle ϕ, which accounts for impedance mismatches. **b)** This further manifests in the asymmetrically distorted Lorentzian amplitude, whose right shoulder exceeds the left one. All the amplitudes are offset corrected by a global factor including attenuation and amplification of the measurement setup and **c)** the phase signal is shifted by the electrical delay and a global phase offset of the setup. The circle fit routine [Pro+15] yields the quality factors $Q_l \approx 51300$, $Q_c \approx 64400$, and consequently $Q_i \approx 250500$.

Power dependence of high-kinetic inductance microwave resonators

Microwave resonators with a substantial kinetic inductance feature significant power dependence. On the one hand, the resonant frequency, given in Eq. (2.36) is dominated by the nonlinear kinetic inductance of Eq. (2.30) and consequently power-dependent, too. Thereby, the relevant scale is the induced current flowing through the resonators with respect to the superconductor's critical current I_c. This nonlinear resonant frequency also affects the whole S_{21} spectra for high powers on the other hand: While sweeping the probe frequency close to and even through the initial resonance, the resonator becomes more and more excited, which in turn increases the kinetic inductance, and thus lowers the effective resonant frequency [Swe+13]. This behavior is often called "bifurcation" and exhibits interrupted circles in the complex plane and abrupt jumps in amplitude and phase, as shown in Fig. 2.6. As a consequence, the S_{21}-parameter that is analytically given in Eq. (2.40) is only valid for small probe powers, meaning few photons in the resonator.

The average photon number is a suitable power scale, especially with regard to quantum systems, interacting with single photons. Therefore, the total energy in the definition of the internal quality factor Q_i of Eq. (2.37) is considered as $E_{tot} = \langle n_{photon} \rangle \hbar\omega^r$. In the notch-type configuration, the power lost in the system

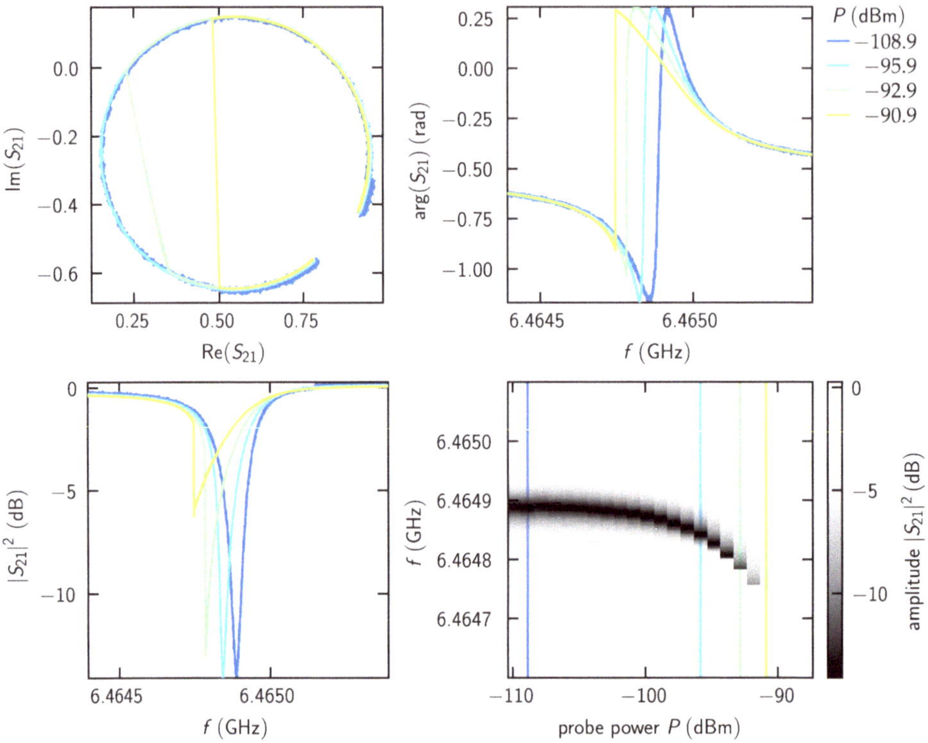

Figure 2.6: Power dependence of high-kinetic inductance microwave resonators. The resonance frequency of devices with dominating nonlinear kinetic inductance strongly depends on the drive power until the resonance vanishes. This entails bifurcating S_{21} spectra, featuring a discrete jump both in amplitude and phase as well as a discontinuous circle in the complex plane. The given powers bear on the feedline to which the notch-type resonator couples.

can further be related to the input power and the reflection and transmission coefficients $P_{\text{loss}} = P_{\text{in}} - P_{\text{out}} = P_{\text{in}}\left(1 - |S_{11}|^2 - |S_{21}|^2\right)$. On resonance ($\omega = \omega^{\text{r}}$), the S-parameters of Eq. (2.40) become $S_{21}(\omega^{\text{r}}) = 1 - S_{11}(\omega^{\text{r}}) = 1 - \frac{Q_{\text{l}}}{Q_{\text{c}}}$ and in consequence, the average photon number in a notch-type resonator holds [Sch20]

$$\langle n_{\text{photon}} \rangle = \frac{2P_{\text{in}}}{\hbar \omega^{\text{r}2}} \frac{Q_{\text{l}}^2}{Q_{\text{c}}}. \tag{2.42}$$

2.3.3 Coupled resonant systems

The load, coupled to the transmission line in notch-type configuration, as illustrated in Fig. 2.3, naturally can not only consist of one microwave resonator, but also of two or even more coupled resonating systems. Such arrangements are of particular interest, if the individual resonant frequencies may be tuned by an

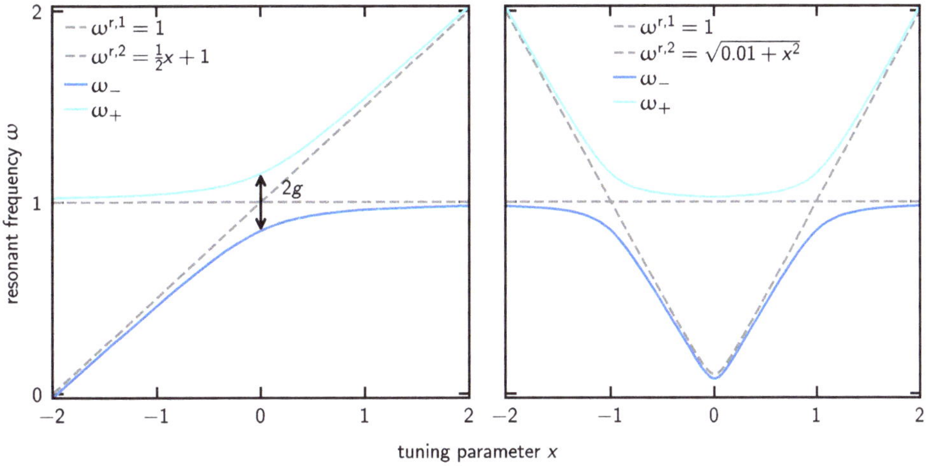

Figure 2.7: Avoided level crossings of two coupled resonant systems with constant coupling strength, but different dispersion relations. If the two individual systems are strongly detuned, they are mostly unaffected by the other, such that their resonant frequencies are shifted only weakly. The closer to bare eigenfrequencies (dashed gray lines) come to each other, the larger the new eigenfrequencies (solid lines) repel each other. Where both bare frequencies would coincide, the frequency difference $\omega_+ - \omega_-$ takes its minimal value of $2g$.

external parameter, as for instance by magnetic or electric fields, such that their bare resonant frequencies would cross each other. However, if the systems can exchange energy at a constant rate g, their oscillation modes will hybridize when their resonance frequencies are close, $\left|\omega^{r,1} - \omega^{r,2}\right| \sim g$. The interplay between these two coupled systems, each oscillating with $a_j(t) = a_j e^{i\omega^{r,j}t}$, is described by the eigenvalue equation [Nol18]

$$\begin{pmatrix} \omega - \omega^{r,1} & g \\ g & \omega - \omega^{r,2} \end{pmatrix} \cdot \begin{pmatrix} a_1(t) \\ a_2(t) \end{pmatrix} = 0 \tag{2.43}$$

and results in the new eigenfrequencies

$$\omega_\pm = \frac{\omega^{r,1} + \omega^{r,2}}{2} \pm \frac{\sqrt{\left(\omega^{r,1} - \omega^{r,2}\right)^2 + 4g^2}}{2}. \tag{2.44}$$

These solutions are schematically depicted in Fig. 2.7 and reveal an avoided level crossing, also called "anticrossing".

The approach of Eq. (2.43) provides merely information about the eigenfrequencies, but not on the whole spectra including all specific linewidths. Within the scope of the input-output theory [GC08], the S_{21}-parameter of a resonant system with bare frequency $\omega^{r,2}$ and internal linewidth $\kappa_{i,2}$, coupled to a notch-type resonator

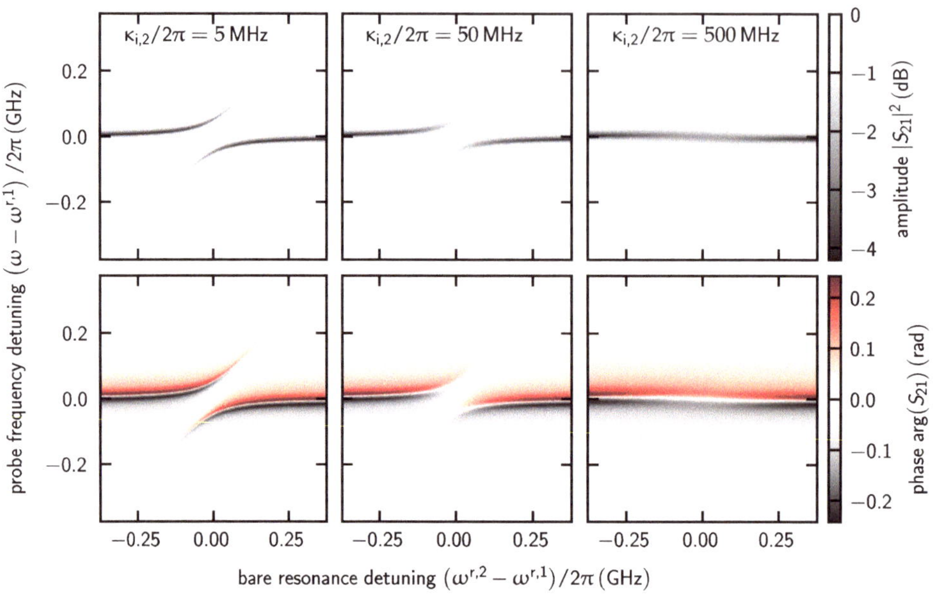

Figure 2.8: Calculated S_{21} spectra of avoided level crossings at different internal quality factors. The first resonator is chosen to be critically coupled, its linewidths are kept fixed at $\kappa_{i,1}/2\pi = \kappa_c/2\pi = 10\,$MHz, and the coupling strength between both systems remains constant at $g/2\pi = 50\,$MHz, too. From left to right, both amplitude and phase signal blur out, during the transition from strong to weak coupling by increasing the internal linewidth of the second resonator $\kappa_{i,2}$.

(coupling constant $2g$) with its resonant frequency $\omega^{r,1}$, internal, and coupling linewidth $\kappa_{i,1}$ and κ_c[10] is given by [Sar+15]

$$S_{21}(\omega) = 1 - \frac{\kappa_c}{2i\left(\omega - \omega^{r,1}\right) + \left(\kappa_c + \kappa_{i,1}\right) + \frac{4g^2}{2i(\omega - \omega^{r,2}) + \kappa_{i,2}}}. \tag{2.45}$$

Both amplitude and phase of the S_{21}-parameter are plotted in Fig. 2.8 depending on the internal linewidth of the second resonator $\kappa_{i,2}$. In the case of strong coupling $\kappa_{i,2} < g$ (left), the two new eigenfrequencies are clearly visible and correspond to Eq. (2.44), but the avoided level crossing becomes more and more indistinct with increasing losses in the second system. In this case of weak coupling, the second resonator is nothing but an additional loss channel for the first one since its excitation decays faster than it can couple back to the first resonator. This mechanism further accounts for the line widening of the first resonator in case of weak coupling (right).

[10] Note that all linewidths κ are defined as full width at half maximum, whereas the coupling constant g is the half width at half maximum for reasons of convention.

2.3.4 Coherence and relaxation properties of a two-level system

In general, not only two harmonic systems, but any kind of resonant system can be coupled to a harmonic oscillator. In particular, superconducting resonators are widely used as detectors for coupled quantum systems and to read out their state. In the simplest case, such a circuit is an effective two-level system comparable to a spin 1/2 particle, having a ground state $|0\rangle = (0,1)^\mathsf{T}$ and an excited state $|1\rangle = (1,0)^\mathsf{T}$ orthogonal to $|0\rangle$. Thus, every possible state of this "quantum bit", often abbreviated as "qubit", can be expressed as a superposition thereof [Kra+19]

$$|\Psi\rangle = a\,|0\rangle + b\,|1\rangle\,. \tag{2.46}$$

Since $|\Psi\rangle$ is normalized to $\langle\Psi\,|\,\Psi\rangle = 1$, the complex amplitudes must hold $|a|^2 + |b|^2 = 1$, where $|a|^2$ and $|b|^2$ describe the probabilities that the system is in the ground and excited state, respectively. The global phase of $|\Psi\rangle$ has no physical meaning, so $a \in \mathbb{R}$ can be chosen to be real-valued. In spherical coordinates with the azimuth angle θ and the polar angle ϕ, an arbitrary state reads [Kra+19; Sch20]

$$|\Psi\rangle = \cos\left(\frac{\theta}{2}\right)|0\rangle + \mathrm{e}^{\mathrm{i}\phi}\sin\left(\frac{\theta}{2}\right)|1\rangle\,. \tag{2.47}$$

and is on the surface of a sphere, the so-called Bloch sphere [Blo46]. For historical reasons, $|0\rangle$ is aligned in positive and $|1\rangle$ in the negative z-direction. In this picture, any qubit operation can be described as linear combination of the Pauli matrices $\hat{\sigma}_x$, $\hat{\sigma}_y$, and $\hat{\sigma}_z$ that correspond to rotations of π around the respective axis. For example, creation and annihilation operators that excite the ground state $\hat{\sigma}_+\,|0\rangle = |1\rangle$ and deexcite the excited state $\hat{\sigma}_-\,|1\rangle = |0\rangle$ can be defined as

$$\hat{\sigma}_\pm = \frac{1}{2}\left(\hat{\sigma}_x \pm \mathrm{i}\hat{\sigma}_y\right). \tag{2.48}$$

In terms of the qubit transition energy $E_{01} = \hbar\omega_{01}$, the qubit Hamiltonian is given by

$$\mathcal{H}_\mathrm{qubit} = \frac{\hbar\omega_{01}}{2}\hat{\sigma}_z. \tag{2.49}$$

The time evolution of the pure qubit state given in Eq. (2.47) is governed by

$$\begin{aligned}|\Psi(t)\rangle &= \hat{\mathcal{U}}(t,t_0)\,|\Psi(t_0=0)\rangle = \exp\left(-\frac{\mathrm{i}}{\hbar}\mathcal{H}_\mathrm{qubit}\cdot(t-t_0)\right)|\Psi(t_0=0)\rangle \\ &= \cos\left(\frac{\theta}{2}\right)|0\rangle + \mathrm{e}^{\mathrm{i}(\phi-\omega_{01}t)}\sin\left(\frac{\theta}{2}\right)|1\rangle\,,\end{aligned} \tag{2.50}$$

using the time evolution operator $\hat{\mathcal{U}}$ and the qubit Hamiltonian of Eq. (2.49). This time evolution corresponds to a continuous rotation around the z-axis with the transition frequency ω_{01}, i.e., a Lamor precession.

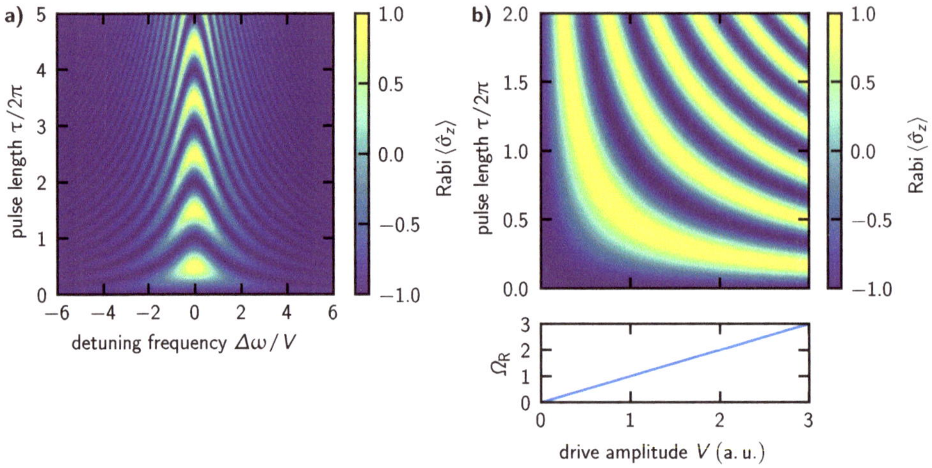

Figure 2.9: Drive frequency and amplitude dependence of Rabi oscillations. The qubit population oscillates with the pulse length of a photon field, where the frequency strongly depends on the drive frequency and amplitude (see Eq. (2.52)). **a)** This Rabi frequency is maximal for zero detuning $\Delta\omega$, meaning the drive frequency ω_d exactly matches the qubit's transition frequency ω_{01}. Small detunings give rise to typical Chevron patterns and the Rabi frequency approaches $\Delta\omega$ for large detunings. **b)** For quantum mechanical Rabi oscillations, their frequency increases linearly with the amplitude of the drive tone.

Rabi oscillations

In order to drive a qubit transition, a microwave tone of amplitude V and frequency ω_d is applied, which couples transversally, and thus in terms of the Pauli matrix $\hat{\sigma}_x$. Therefore, the Hamiltonian of a driven qubit holds [SN17; Bra18; Sch20]

$$\mathcal{H} = \frac{\hbar\omega_{01}}{2}\hat{\sigma}_z + \hbar V\cos(\omega_d t)\hat{\sigma}_x. \tag{2.51a}$$

As the microwave drive itself rotates with the ω_d around the z-axis, the qubit is more intuitively described within the rotating frame of this electromagnetic field. Using the unitary operation $\hat{U} = \exp(\mathrm{i}\omega_d t\hat{\sigma}_z/2)$, the driven qubit Hamiltonian of Eq. (2.51a) can be transformed from the reference to the rotating frame and results in

$$\tilde{\mathcal{H}} = \frac{\hbar\Delta\omega}{2}\hat{\sigma}_z + \frac{\hbar V}{2}\hat{\sigma}_x. \tag{2.51b}$$

Here, only rotations of the frequency detuning $\Delta\omega = \omega_{01} - \omega_d$ remain. The diagonalized eigenfrequency corresponds to the generalized Rabi frequency $\Omega_R = \sqrt{V^2 + \Delta\omega^2}$ and depends on the frequency detuning $\Delta\omega = \omega_{01} - \omega_d$ and the drive amplitude V.

Analysis of the time evolution in the new basis yields the qubit's excitation probability $p_{|1\rangle}$, or rather the expectation value of $\hat{\sigma}_z$ [SN17]

$$p_{|1\rangle} = \frac{1}{2}\left(1 - \left(\frac{\Delta\omega}{\Omega_\mathrm{R}}\right)^2\right)(1 - \cos(\Omega_\mathrm{R} t)), \qquad \langle\hat{\sigma}_z\rangle = 2p_{|1\rangle} - 1. \tag{2.52a}$$

Dispersive readout

Nowadays, qubits are often manipulated and read out via the resonator to which they are coupled. In analogy to an atom in a cavity as in optics, the resonator-qubit interaction can be described by the Jaynes-Cummings model. There, the total Hamiltonian $\mathcal{H}_\mathrm{JC}$ composes of parts of the bare harmonic oscillator $\mathcal{H}_\mathrm{r}$, the pure qubit $\mathcal{H}_\mathrm{qubit}$, and their interaction $\mathcal{H}_\mathrm{int}$ and reads in the rotating frame [JC63; HR06; WM08].

$$\mathcal{H}_\mathrm{JC} = \underbrace{\hbar\omega_\mathrm{r}\left(\hat{a}^\dagger\hat{a} + \frac{1}{2}\right)}_{\mathcal{H}_\mathrm{r}} + \underbrace{\frac{\hbar\omega_{01}}{2}\hat{\sigma}_z}_{\mathcal{H}_\mathrm{qubit}} + \underbrace{\hbar g\left(\hat{\sigma}_+\hat{a} + \hat{\sigma}_-\hat{a}^\dagger\right)}_{\mathcal{H}_\mathrm{int}}. \tag{2.53}$$

Here, $\hat{a}^\dagger$ and $\hat{a}$ denote the photon creation and annihilation operator, respectively, ω_r the resonator's harmonic eigenfrequency, and g the coupling strength between resonator and qubit.

In general, qubit and resonator exchange energy, which is why the eigenstates of the coupled system are neither eigenstates of the pure resonator nor of the qubit. This means that a part of the resonator excitation is in the qubit and vice versa, enabling manipulation, readout, and decay of the qubit via the resonator.

If the detuning of the resonator and the qubit is large compared to their coupling $\Delta\omega = \omega_\mathrm{r} - \omega_{01} \gg g$, the eigenstates are similar to those of the pure resonator and qubit. In this so-called dispersive limit [Bla+04], the system's eigenstates can be approximated as product states of resonator and qubit, while the coupling is only a small perturbation. Perturbation analysis of the Jaynes-Cummings Hamiltonian (2.53) yields

$$\tilde{\mathcal{H}}_\mathrm{JC} = \hbar\left(\omega_\mathrm{r} + \chi\hat{\sigma}_z\right)\hat{a}^\dagger\hat{a} + \frac{1}{2}\left(\omega_{01} + \chi\right)\hat{\sigma}_z + \mathcal{O}\left(\frac{g^2}{\Delta\omega^2}\right) \tag{2.54}$$

with the dispersive shift $\chi = g^2/\Delta\omega$ for true two-level systems without higher harmonics. Since the Jaynes-Cummings Hamiltonian in the dispersive limit (Eq. (2.54)) is diagonalized, the effective qubit $\tilde{\omega}_{01} = \omega_{01} + \chi$ and resonator frequencies $\tilde{\omega}_\mathrm{r} = \omega_\mathrm{r} + \chi\hat{\sigma}_z$ equal the diagonal elements. The latter's dependence on $\hat{\sigma}_z$ means that the qubit state can be determined by measuring the resonator frequency. This, however, is a projective measurement, where the qubit's wave function (2.47) collapses either into $|0\rangle$ or $|1\rangle$ with the probabilities $\cos^2(\theta/2)$ and $\sin^2(\theta/2)$, so only $\tilde{\omega}_r = \omega_r \pm \chi$ can be measured. Repeating the same measurement many times to perform statistics yields the expectation value of the qubit state $\langle\hat{\sigma}_z\rangle$ that corresponds to the qubit's excitation probability as [Sch20]

$$\langle\hat{\sigma}_z\rangle = \langle\Psi\,|\,\hat{\sigma}_z\,|\,\Psi\rangle = -|a|^2 + |b|^2 = -\cos\theta. \tag{2.55}$$

2.4 Dynamics of long Josephson junctions

2.4.1 Basics of the Josephson effect

Key elements of manifold superconducting devices, such as superconducting quantum bits [NPT99; Wal+00a; CW08], but also superconducting quantum interference devices (SQUIDs) [Jak+64; ZTH70; Mer70; Cla66], and voltage standard circuits [Tay+67; FFT73; Ham00], are Josephson tunnel junctions as they feature unique electromagnetic characteristics. The Josephson effect was predicted by Brian David Josephson in 1962 [Jos62] and experimentally proven for the first time by Philip Warren Anderson and John Martin Rowell in 1963 [AR63].

It describes coherent Cooper pair tunneling between weak link superconductors, separated by a thin insulating or normal conducting barrier or constrictions, point contacts, etc. Due to the good controllability of electronic properties, superconductor-insulator-superconductor (SIS) junctions find widespread usage and are therefore utilized in this work. It is remarkable for tunneling processes between superconductors that the tunneling probability of a whole Cooper pair equals the one of single electrons, and thus is of measurable order of magnitude. This originates in the superconductors' macroscopic quantum states, whose wave functions are reduced from the bulk values inside the tunnel barrier but overlap nonvanishing. In a perturbation theory treatment, this manifests by a constant coupling energy g on the off-diagonal of the time-dependent Schrödinger equation [FLS64a]

$$\mathrm{i}\hbar\partial_t \begin{pmatrix}\psi_1\\ \psi_2\end{pmatrix} = \begin{pmatrix}\mathcal{H}_1 & g\\ g & \mathcal{H}_2\end{pmatrix}\begin{pmatrix}\psi_1\\ \psi_2\end{pmatrix}. \tag{2.56}$$

Thereby, both superconductors hold the macroscopic wave function $\Psi_j(t) = \sqrt{n_{sj}(t)}\mathrm{e}^{\mathrm{i}\theta_j(t)}$, as given in Eq. (2.3) that solves the individual Schrödinger equation $\mathrm{i}\hbar\partial_t\psi_j = \mathcal{H}_j\psi_j = E_j\psi_j$ with the ground state eigenenergies E_j. Assuming weak tunneling between identical superconductors with $n_{s1} = n_{s2}$ and $\dot{n}_{s1} = -\dot{n}_{s2}$, the real part of Eq. (2.56) results in the so-called first Josephson equation

$$j_J = j_c \sin(\varphi). \tag{2.57}$$

Instead of a DC voltage, as expected in case of normal state tunneling, Eq. (2.57) reveals that DC supercurrents across a tunnel barrier need to be driven by the gauge-invariant phase difference

$$\varphi = \theta_2 - \theta_1 - \frac{2\pi}{\Phi_0} \int_1^2 \mathbf{A} d\mathbf{l} \tag{2.58}$$

up to the junction's critical current density across the barrier $j_c = 2gn_s/\hbar$. This quantity is a substantial parameter and includes information about barrier properties, temperature dependence, etc.

Within the scope of the BCS theory, Vinay Ambegaokar and Alexis Baratoff gave a description of coherent Cooper pair tunneling between two, in general different, superconductors [AB63; Wal76; BP82; Wol11]. They found the dependency of the critical current of lumped element junctions on the normal state tunnel resistance R_n and the gap energies Δ_1 and Δ_2 as

$$I_c = \frac{\Delta_1(T)}{eR_n} K\left(\sqrt{1 - \left(\frac{\Delta_1(T)}{\Delta_2(T)}\right)^2}\right). \tag{2.59a}$$

Here, $K(x)$ denotes the complete elliptic integral of the first kind and in case of identical superconductors, Eq. (2.59a) simplifies to the well-known Ambegaokar-Baratoff relation

$$I_c = \frac{\pi\Delta(T)}{2eR_n} \tanh\left(\frac{\Delta(T)}{2k_B T}\right) \overset{T\to 0}{\longrightarrow} \frac{\pi\Delta(T=0)}{2eR_n} \overset{(2.25)}{\approx} 2.388 \times 10^{-4}\,\mathrm{V\,K^{-1}} \cdot \frac{T_c}{R_n}. \tag{2.59b}$$

Hence, the critical current of the Josephson junction can be estimated by the normal conducting tunnel resistance R_n, the critical temperature T_c or the gap energy Δ and allows experimentalists to characterize the tunnel junctions at room temperature without the elaborate cooling far below T_c. From the imaginary part of Eq. (2.56) follows another characteristic dependency, commonly referred to as the second Josephson equation. If the eigenenergies of the two weak link superconductors are shifted against each other by $E_2 - E_1 = 2eV$, this difference in the electrical potentials is given by

$$V(t) = \frac{\hbar}{2e}\frac{d\varphi}{dt} = \frac{\Phi_0}{2\pi}\frac{d\varphi}{dt}. \tag{2.60}$$

According to the two Josephson equations, applied constant voltages $V < 2\Delta/e$ imply a rotating phase difference, and thus an oscillating current across the barrier, often referred to as the AC Josephson effect. Moreover, the second Josephson equation resembles Faraday's law of inductance with the phase-flux relation $\varphi = \frac{2\pi}{\Phi_0}\Phi$. However, the voltage does not originate in the magnetic energy but in the Cooper pair's kinetic energy instead. In analog to Sec. 2.2.1, this goes along with a inherent kinetic-type inductance [Jos64]

$$L(\varphi) = \frac{V}{\frac{\partial I(\varphi)}{\partial t}} = \frac{V}{\frac{\partial I}{\partial \varphi}\frac{\partial \varphi}{\partial t}} = \frac{V}{I_c \cos(\varphi)\frac{2\pi}{\Phi_0}V} = \frac{\Phi_0}{2\pi I_c \cos(\varphi)} = \frac{L_J}{\cos(\varphi)}. \tag{2.61}$$

Depending on the phase difference applied, and thus the current flowing across the tunnel barrier, Josephson junctions exhibit a nonlinear inductance with the minimum absolute value $L_J \equiv \frac{\hbar}{2eI_c} = \frac{\Phi_0}{2\pi I_c}$, the so-called Josephson inductance [Wal76]. This unique nonlinear inductance makes Josephson junctions key elements for most superconducting quantum circuits since it enables, for instance, anharmonic LC oscillators with distinct artificial two-level systems utilized as quantum bits. The AC Josephson effect, originating in the nonlinear Josephson inductance implies, that energy is stored in the lumped element junction. When increasing a bias current from zero (where $\varphi = 0$, too) to a finite value $I < I_c$, the phase difference has to change consistent with the first Josephson equation (2.57). Although the voltage will be zero, when reaching the current I and Cooper pairs can tunnel across the junction without any dissipation, while this phase change yields a nonzero voltage during the current increase. As a consequence, an external current source has to do work on the junction and the free energy stored in the kinetic energy of the moving superfluid is given by [OD91; GMD16]

$$\mathcal{F}_J = \int_0^t V I \mathrm{d}\tilde{t} = \int_0^t \frac{\Phi_0}{2\pi}\frac{\mathrm{d}\tilde{\varphi}}{\mathrm{d}\tilde{t}} I_c \sin(\tilde{\varphi})\mathrm{d}\tilde{t} = \frac{\Phi_0 I_c}{2\pi}\int_0^{\varphi} \sin(\tilde{\varphi})\mathrm{d}\tilde{\varphi} = E_J\left(1-\cos(\varphi)\right). \tag{2.62}$$

From another perspective, this energy resembles molecular binding energy between atoms, originating in the overlap of the macroscopic wave functions and the increased accessible space for Cooper pairs. Therefore, $E_J \equiv \frac{\Phi_0 I_c}{2\pi}$ is referred to as Josephson coupling energy [Jos65; BP82].

Resistively and capacitively shunted junction model for current biased junctions

In order to gain an insight into the phase dynamics of lumped element Josephson junctions, a useful model was established by Wilbert C. Stewart and Dean E. McCumber [Ste68; McC68]. On the one hand, the Josephson junction's weak link can be considered as two conductors, separated by a dielectric layer, like in a capacitor.[11] On the other hand, the weak link also provides tunneling of both superconducting Cooper pairs and (in case of finite temperature or magnetic field) normal conducting quasiparticles like a tunnel junction. As a consequence, a Josephson junction can be considered as a parallel connection of a resistor with current $I_R = \frac{V}{R}$, accounting for ohmic quasiparticle tunneling, a capacitor with its displacement current $I_C = \dot{q} = C\frac{\mathrm{d}V}{\mathrm{d}t}$ and the Josephson junction with $I_\mathrm{J} = I_\mathrm{c} \sin\varphi$, representing Cooper pair tunneling. For such resistively and capacitively shunted junctions (RCSJ) [McC68], [Ste68], the net current reads

$$I = C\frac{\mathrm{d}V}{\mathrm{d}t} + \frac{V}{R} + I_\mathrm{c}\sin\varphi \overset{(2.60)}{=} \frac{\Phi_0 C}{2\pi}\frac{\mathrm{d}^2\varphi}{\mathrm{d}t^2} + \frac{\Phi_0}{2\pi R}\frac{\mathrm{d}\varphi}{\mathrm{d}t} + I_\mathrm{c}\sin\varphi. \tag{2.63a}$$

In dimensionless units Eq. (2.63a) can be rewritten as

$$\gamma = \frac{\mathrm{d}^2\varphi}{\mathrm{d}\bar{t}^2} + \frac{1}{\sqrt{\beta_\mathrm{c}}}\frac{\mathrm{d}\varphi}{\mathrm{d}\bar{t}} + \sin\varphi, \tag{2.63b}$$

where $\gamma = \frac{I}{I_\mathrm{c}}$ denotes the normalized bias current and the time is normalized to $\bar{t} \equiv \omega_\mathrm{p} t$ in terms of the inverse plasma frequency of the junction ω_p.[12] This characteristic frequency describes the frequency of small phase oscillations at zero bias current and equals the resonance frequency of the LC-circuit formed by the Josephson inductance L_J and the junction capacitance C

$$\omega_\mathrm{p} \equiv \sqrt{\frac{2\pi I_\mathrm{c}}{\Phi_0 C}} = \frac{1}{\sqrt{L_\mathrm{J} C}}. \tag{2.64}$$

Moreover, the so-called Stewart-McCumber parameter $\beta_\mathrm{c} \equiv \frac{2\pi}{\Phi_0} R^2 C I_\mathrm{c}$ is introduced as a measure for damping. The dissipation of a Josephson junction is caused by ohmic quasiparticle tunneling, whose corresponding time constant equals $\tau_{RC} = RC$. Hence, the quality factor of a Josephson junction is given by [Tin04]

$$Q = \omega_\mathrm{p}\tau_{RC} = \sqrt{\beta_\mathrm{c}}. \tag{2.65}$$

Equation (2.63) can be considered as a classical equation of motion for a fictitious phase particle of mass $m_\varphi = C(\Phi_0/2\pi)^2$ and is well comparable with a damped gravity pendulum, described by $m\ddot{x} - \zeta\dot{x} = -\frac{\partial U}{\partial x}$. In analogue, the potential energy of a current biased Josephson junction in the RCSJ model can be identified as

$$U(\varphi) = -E_\mathrm{J}\left(\gamma\varphi + \cos\varphi\right), \tag{2.66}$$

[11] The existence of a junction capacitance is in general not necessarily given (for instance in superconductor-normalconductor-superconductor junctions), but the resistively and capacitively shunted junction model is an adequate description of dielectric tunnel barriers such as in SIS junctions used throughout this work.

[12] The term "plasma" frequency is due to its dispersion relation, which equals plasma oscillations in metals, as can be seen later in Sec. 2.4.3.

which is often called "tilted washboard potential" and is illustrated in Fig. 2.10 a). Without any bias current γ, the potential energy is cosinusoidal, but applying bias currents tilts the potential proportionally to γ. This yields local maxima at $\varphi_b = \pi - \arcsin(\gamma)$ and local minima at $\varphi_0 = \arcsin(\gamma)$, giving rise to metastable states of the phase particle. The barrier height in between is

$$U_0 = U(\varphi_b) - U(\varphi_0) = 2E_J\left(\sqrt{1-\gamma^2} - \gamma \arccos(\gamma)\right) \tag{2.67a}$$

$$\approx \frac{4\sqrt{2}}{3}E_J(1-\gamma)^{\frac{3}{2}} + \frac{\sqrt{2}}{15}E_J(1-\gamma)^{\frac{5}{2}} + \mathcal{O}\left((1-\gamma)^3\right). \tag{2.67b}$$

The typical experimental situation $E_J \gg k_B T$ implies that thermal excitation becomes relevant for $\gamma \to 1$, the center of the Taylor expansion, given in Eq. (2.67b).

Due to its bias current dependence, the potential wells are in general anharmonic and current-dependent, too. By Taylor expanding the potential around the local minima, the frequency of small amplitude oscillations is related to the potential curvature in the well $\left.\frac{\partial^2 U(\phi)}{\partial \phi^2}\right|_{\phi=\phi_0}$ and the oscillating particle's mass m_φ by [Dah+68]

$$\omega_0(\gamma) = \sqrt{\frac{\left.\frac{\partial^2 U(\varphi)}{\partial \varphi^2}\right|_{\varphi=\varphi_0}}{m_\varphi}} = \omega_p\left(1-\gamma^2\right)^{1/4}. \tag{2.68}$$

Beyond these energetic and dynamic considerations, the RCSJ model provides information about Josephson junction's time averaged IV characteristics (IVC). Within the scope of the tilted washboard potential, the IVC shows a vertical slope in the superconducting branch below I_c due to $\langle\dot{\varphi}\rangle = \langle V\rangle = 0$ (zero-voltage state) and a finite slope R_n^{-1} in the normal conducting branch above I_c, where the current cannot be carried completely by the supercurrent through the junction but also by dissipative quasiparticles.

The IVCs vary for different Stewart-McCumber parameters, as can be seen in Fig. 2.10 b). In the overdamped case $\beta_c \ll 1$, even though the phase particle can move from one valley to another for $I_b \gtrsim I_c$, it does not achieve the running state. This is because the released energy is dissipated, and the phase particle needs to be activated in every single valley. Hence, the rate $\frac{d\varphi}{dt}$ is small and the resulting voltage does not jump directly to the gap voltage $V_\Delta = \frac{2\Delta}{e}$. In the underdamped case $\beta_c \gg 1$, however, the phase particle can run down the full potential after overcoming the first potential well since the gained energy (minus the dissipated energy that is small because of low damping) suffices to overcome the next potential well. As a consequence, this running state maintains, even if the bias current I_b

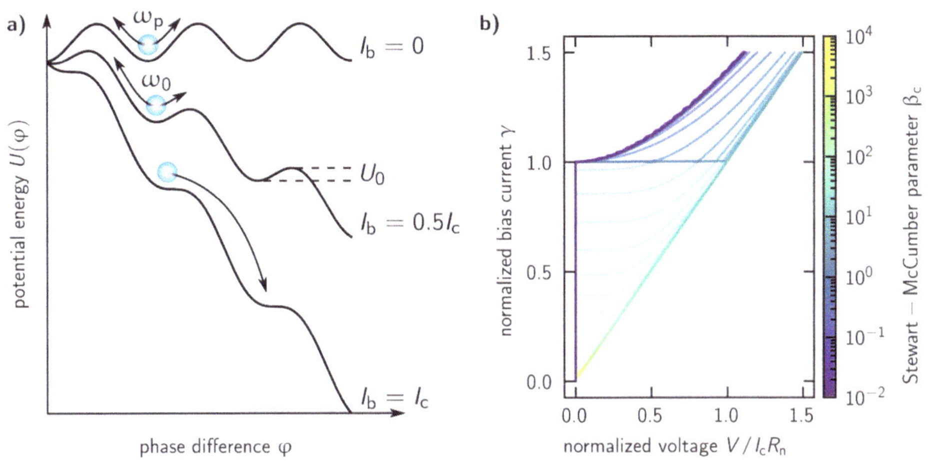

Figure 2.10: Potential energy for varying bias currents and numerically simulated IV-characteristic for different damping values within the scope of the RCSJ model. a) For zero bias current, the phase particle is trapped in a minimum of the potential well with potential height E_J and oscillates inside with the specific plasma frequency ω_p. Since ω_p is typically in the order of tens to hundreds of gigahertz, in transport measurements, only time averaged values are measured. These mean values are time-independent $\langle\dot{\varphi}\rangle = 0$ and according to the second Josephson equation (2.60) no voltage drops across the junction. However, an applied bias current tilts the washboard potential and if it exceeds the critical current $I_b > I_c$ the phase particle can overcome the potential well with a finite phase velocity $\langle\dot{\varphi}\rangle \neq 0$ which results in a voltage drop across the junction. **b)** A damping-independent superconducting behavior characterizes the up-sweep of the bias current I_b up to the critical current I_c, where the voltage drops abruptly to the superconducting gap voltage V_Δ in case of low damping (dark line color). Beyond an underdamped junction shows an ohmic dependency due to ohmic quasiparticle tunneling. With increasing damping (brighter line color), this mentioned voltage drop decreases, and normal conducting behavior approaches the linear ohmic slope slowly and more slowly. Furthermore, the hysteretic behavior between up- and down-sweep of the bias current increases with decreasing damping (from dark blue to yellow), meaning the IVC follows the ohmic straight below I_c down to a retrapping current I_r, where it becomes superconducting again.

is decreased again down to a retrapping current $I_r < I_c$. The resulting hysteresis and, in particular, the ratio of retrapping to critical current is a measure of the Josephson junction's quality factor, given in Eq. (2.65).

Although the RCSJ model describes significant parts of the junction's IV characteristics correctly, such as the occurrence of a zero-voltage state and a damping-dependent hysteresis, it is essential to recognize that real Josephson junctions do not follow the ohmic branch during the down-sweep, but remain almost constant at the gap voltage V_Δ (see Fig. 2.11 a)). At this branch, the RCSJ model is limited by its assumption of an ohmic, voltage-independent quasiparticle tunneling resistance.

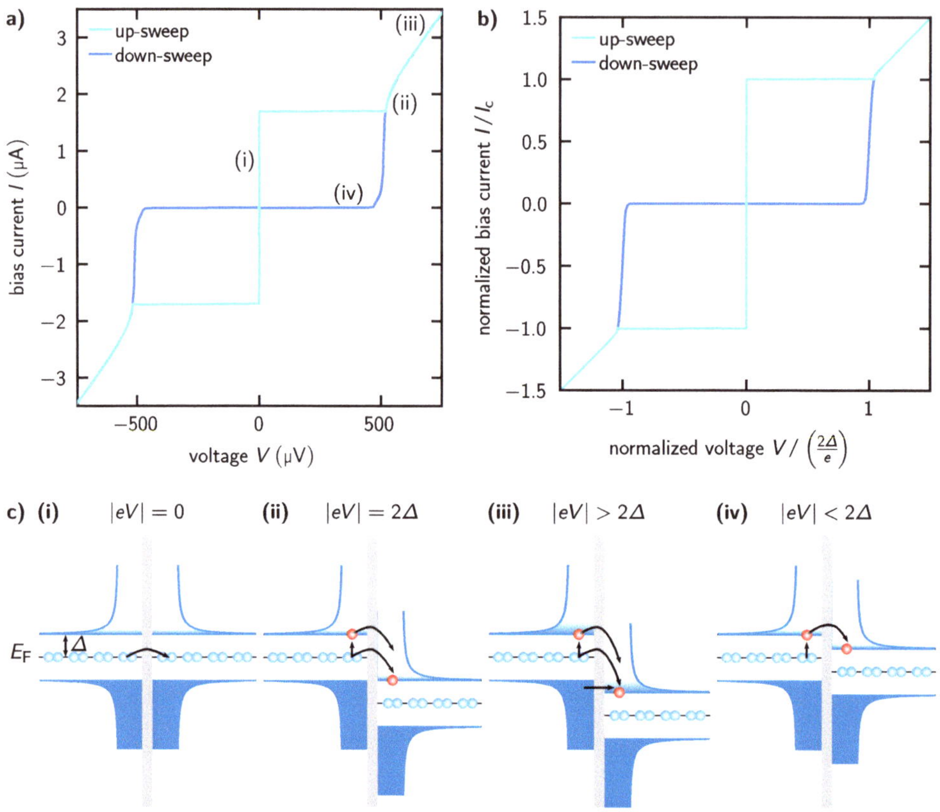

Figure 2.11: IV-characteristic of a current biased and underdamped Josephson junction. a) IVC of a 110 μm × 5 μm wide underdamped Josephson junction with electrodes made from AlO_x and pure Al measured at $T \approx 20\,\text{mK}$. **b)** The simulated IVC by means of the nRCSJ model in eq (2.69) exhibits an almost vertical slope (conductance) around the gap voltage at the down-sweep between the critical and the retrapping current. **c)** Various conductance on different branches of a real junction's IVC within the scope of band diagrams of two weak link superconductors [Poo+14]. **(i)** Below the critical current, the junction remains in the zero-voltage state, where superconducting (s) Cooper pairs may tunnel coherently and without any dissipation (s → s). **(ii)** At the critical current, the junction jumps to the gap voltage $2\Delta/e$, while Cooper pairs are broken in two quasiparticles (q) each. These fermions may tunnel through the barrier dissipatively with a nonlinear resistance due to the quasiparticles' increased density of states (s → q). **(iii)** Far above the critical current, and thus way beyond the gap voltage, lots of normal free electrons (n) tunnel yielding the ohmic normal state resistance (n → n). When decreasing the bias below the critical current underdamped junctions remain in the running state at the gap voltage down to a retrapping current (analog to **(ii)**). **(iv)** Close to the retrapping current, and thus below the gap voltage, a small number of not yet recondensed quasiparticles may tunnel with dissipation (q → q), which is often referred to as Stewart-McCumber branch.

In real Josephson junctions, however, this resistance is much smaller for voltages close to the gap voltage V_Δ than for those far above. This nonlinear resistance around the gap originates in the drastically increased quasi-particle density (see Fig. 2.2) that are available for only few thermally activated quasiparticles [Sco70; PS73; BP82; KW19]. This behavior is covered by a nonlinear resistive and capacitively shunted junction (nRCSJ) and can be modeled by the power law expression [MK83]

$$I(V) = \frac{V}{R} \cdot \frac{\left(\frac{V}{V_\Delta}\right)^n}{1 + \left(\frac{V}{V_\Delta}\right)^n}, \qquad n \gg 1 \tag{2.69}$$

and results in a nearly constant gap voltage in the down-sweep, similar to experimental data (compare Fig. 2.11 a) and 2.11 b)).

2.4.2 Magnetic field and size effects of distributed junctions

For all the former mentioned phenomena of Josephson junctions, they were so far taken as point like "lumped elements" with spatially homogeneously distributed supercurrent densities $j_s(x) = j_s$ and phase differences $\varphi(x) = \varphi$. Real SIS junctions, however, involve a spatial extension along x and y of length ℓ and width w, respectively, which may feature spatially varying $j_s(x, y)$ and $\varphi(x, y)$.

Low-inductance limit of bulk electrodes

In order to work out the phase difference introduced between two positions, say, along the x-coordinate, a cross-section of a distributed junction with bulk electrodes, as sketched in Fig. 2.12, is considered. In this scheme, two closely spaced pairs of points in the vicinity of the junction are regarded and the closed contour C through these four points can be split into the two paths across the junction and these along the electrodes

$$\Phi = \oint_C \mathbf{A}\mathrm{d}\mathbf{l} = \underbrace{\int_{r_1^t}^{r_1^b} \mathbf{A}\mathrm{d}\mathbf{l}}_{\text{junction 1}} + \underbrace{\int_{r_1^b}^{r_2^b} \mathbf{A}\mathrm{d}\mathbf{l}}_{\text{bottom el.}} + \underbrace{\int_{r_2^b}^{r_2^t} \mathbf{A}\mathrm{d}\mathbf{l}}_{\text{junction 2}} + \underbrace{\int_{r_2^t}^{r_1^t} \mathbf{A}\mathrm{d}\mathbf{l}}_{\text{top el.}} . \tag{2.70a}$$

Well inside bulk electrodes $d \gg \lambda_L$ screening supercurrents $\mathbf{j}$ are negligible small and the first London equation reduces to $\mathbf{A} = \frac{\Phi_0}{2\pi}\nabla\theta$. Consequently, the phase drops along the bottom and top electrode are described by

$$\frac{2\pi}{\Phi_0} \int_{r_1^b \backslash r_2^t}^{r_2^b \backslash r_1^t} \mathbf{A}\mathrm{d}\mathbf{l} = \theta\left(r_2^b \backslash r_1^t\right) - \theta\left(r_1^b \backslash r_2^t\right). \tag{2.70b}$$

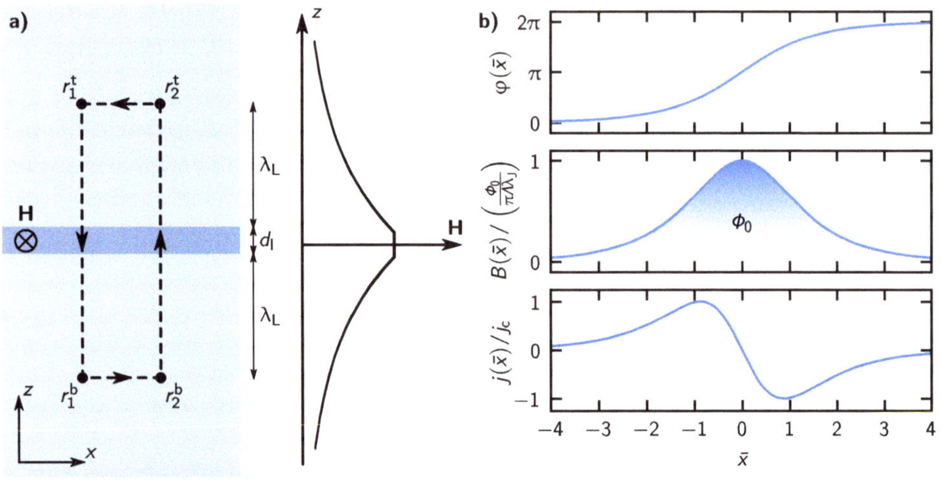

Figure 2.12: Spatial phase variation in distributed junctions and the Josephson vortex solution. a) The Josephson junction is supposed to consist of both a bulk bottom and top electrode that are separated along the xy-plane by a tunnel barrier layer of thickness d_I. An external magnetic field is applied along the y-axis and penetrates the bulk electrodes on the scale of the London penetration depth λ_L. **b)** Gauge-invariant phase difference, magnetic flux density, and current density of the Josephson vortex solution for the Ferrell-Prange equation. The magnetic field and the Josephson current density decay with the characteristic length scale λ_J. The current's opposite signs indicate the ring current associated with 2π phase kink, whereas the net current is zero. The integration of the magnetic flux density B across the vortex's cross-section equals a total flux of the superconducting magnetic flux quantum Φ_0.

With the definition of the Josephson junction's gauge-invariant phase difference in Eq. (2.58), the phase drops across the junctions result in

$$\frac{2\pi}{\Phi_0}\int_{r_1^t\backslash r_2^b}^{r_1^b\backslash r_2^t} \mathbf{A}\mathrm{d}\mathbf{l} = \theta\left(r_1^b\backslash r_2^t\right) - \theta\left(r_1^t\backslash r_2^b\right) \mp \varphi(x_1\backslash x_2), \tag{2.70c}$$

depending on the direction of walking the contour. The enclosed magnetic flux through this the selected area is $\Phi = \mu_0 H_y \cdot (x_2 - x_1)\,\Lambda$ with the magnetic thickness $\Lambda = \lambda_{L,1} + \lambda_{L,2} + d_I$ [Jos64]. For an infinitesimal small spacing $\mathrm{d}x$ in the x-direction, the magnetic field results in

$$H_y(x) = \frac{\Phi_0}{2\pi\mu_0\Lambda}\lim_{x_2\to x_1}\frac{\varphi(x_2)-\varphi(x_1)}{x_2-x_1} = \frac{\Phi_0}{2\pi\mu_0\Lambda}\frac{\mathrm{d}\varphi(x)}{\mathrm{d}x} \propto \frac{\mathrm{d}\varphi}{\mathrm{d}x}. \tag{2.71}$$

The supercurrent along z, driven by this phase gradient, is on the one hand linked with the magnetic field by Ampere's law $j_z = \frac{\mathrm{d}H_y(x)}{\mathrm{d}x}$ and on the other hand defined by the first Josephson equation (2.57) $j_z(x) = j_c \sin(\varphi(x))$. In combination, the phase holds a stationary, unperturbed sine-Gordon equation that is often referred to as Ferrell-Prange equation [FP63]

$$\frac{\Phi_0}{2\pi\mu_0\Lambda j_c}\frac{\mathrm{d}^2\varphi}{\mathrm{d}x^2} = \frac{\mathrm{d}^2\varphi}{\mathrm{d}\bar{x}^2} = \sin(\varphi). \tag{2.72}$$

With the electrode's sheet inductance $L_0^\square = \mu_0 \Lambda$ [Swi61; Lik86], the junction's characteristic length scale is given by

$$\lambda_\mathrm{J} \equiv \sqrt{\frac{\Phi_0}{2\pi\mu_0\Lambda j_\mathrm{c}}} = \sqrt{\frac{\Phi_0}{2\pi L_0^\square j_\mathrm{c}}} \tag{2.73}$$

that is often called as Josephson penetration depth. In dimensionless units, the length x is normalized to λ_J so that $\bar{x} \equiv x/\lambda_\mathrm{J}$. The general solution of this Ferrell-Prange equation is for instance given in [OS67; LS67]. From a physical perspective, the most interesting particular solution in linear geometry junctions in the limit $\ell \to \infty$ are Josephson vortices[13]

$$\varphi_\mathrm{F}(\bar{x}) = \pm 4 \arctan\left(\mathrm{e}^{\bar{x}-\bar{x}_0}\right) + 2\pi n. \tag{2.74a}$$

whose center of mass is located at $\bar{x}_0$. As can be seen in Fig. 2.12 b), this solution is associated with a change of the phase difference across the junction of 2π, which is why it is often referred to as "2π-kink solution". This twist extends over the characteristic length scale of the Josephson penetration length and is centered around $\bar{x}_0$.

From Eq. (2.71), the magnetic flux density that goes along to this solution results in

$$B_{\mathrm{F}y}(\bar{x}) = \frac{\Phi_0}{2\pi\Lambda}\frac{\partial\varphi_\mathrm{F}}{\partial\bar{x}}\frac{\partial\bar{x}}{\partial x} = \pm\frac{\Phi_0}{\pi\Lambda\lambda_\mathrm{J}}\frac{1}{\cosh(\bar{x}-\bar{x}_0)}. \tag{2.74b}$$

The corresponding magnetic flux equals the superconducting magnetic flux quantum $\Phi_\mathrm{F} = \int_{-\infty}^{\infty}\mathrm{d}x\int_{-\Lambda/2}^{\Lambda/2}\mathrm{d}z B_{\mathrm{F}y}(\bar{x}) = \Phi_0$ (see Fig. 2.12 b)), which is why this solution is also often called "fluxon".

According to Ampère's law, the associated supercurrent

$$j_{\mathrm{F}z}(\bar{x}) = \frac{1}{\mu_0}\frac{\partial B_{\mathrm{F}y}}{\partial\bar{x}}\frac{\partial\bar{x}}{\partial x} = \pm\frac{\Phi_0}{\pi\mu_0\Lambda\lambda_\mathrm{J}^2}\frac{\sinh(\bar{x}-\bar{x}_0)}{\cosh^2(\bar{x}-\bar{x}_0)} = 2j_\mathrm{c}\frac{\sinh(\bar{x}-\bar{x}_0)}{\cosh^2(\bar{x}-\bar{x}_0)} \tag{2.74c}$$

comes as a ring current around the tunnel barrier, which is why the denotation "Josephson vortex" solution. It should be noted that in contrast to Abrikosov vortices in type-II superconductors, Josephson vortices feature no normal conducting core. As superconductivity is weakest in the junction region, the fluxon's energy is the lowest around the junction, which is why they enter there and are pinned to the tunnel barrier (see Fig. 2.16).

[13] Richard A. Ferrell, and Richard E. Prange quoted the solution $\varphi(\bar{x}) = 2\arcsin[\mathrm{sech}(\bar{x}-\bar{x}_0)]$ that is an equivalent representation for a semi-infinite Josephson junction ($0 \leq \bar{x} \leq \infty$) which fulfills the boundary condition $\varphi(\infty) = 0$ [FP63].

High-kinetic inductance limit of thin-film electrodes

In Josephson junctions with electrodes made from high-kinetic inductance superconductors, the magnetic penetration depth $\lambda_L \sim 3\,\mu m$ exceeds the typical film thickness $d \sim 20\,nm$, which yields incomplete screening of applied magnetic fields. Therefore, the assumption of vanishing screening currents well inside the electrode, made in Eq. (2.70b) for the derivation of the Ferrell-Prange equation, is no longer valid, and this case needs to be discussed separately.

By solving the Poisson equation of screening currents (2.4b) (see London theory in Sec. 2.1.1) under the boundary conditions of an extended Josephson junction by means of Fourier transformation, Alfimov and Popkov worked out that the phase difference across a junction with thin-film electrodes is given by [IS90; AP95]

$$\sin(\varphi) = \frac{1}{\bar{\lambda}} \int_{-\infty}^{\infty} G_{\bar{d}}\left(\frac{|\bar{x}-u|}{\bar{\lambda}}\right) \frac{d^2\varphi}{du^2} du \tag{2.75a}$$

$$G_{\bar{d}}(v) = \frac{1}{2\pi} \int_{-\infty}^{\infty} \frac{\tanh\left(\bar{d}\sqrt{1+k^2}\right)}{\sqrt{1+k^2}} e^{ikv} dk. \tag{2.75b}$$

Here, $\bar{d} = \frac{d}{\lambda_L}$ denotes the normalized electrode thickness with respect to the London penetration depth and $\bar{\lambda} = \frac{\lambda_L}{\lambda_J^b}$ the ratio of London penetration depth and the bulk Josephson length. In the limit of large λ_J, meaning $\bar{\lambda} \ll 1$, as it is the case throughout this work, the kernel of the integral operator, given in Eq. (2.75b) simplifies to $\lim_{\bar{\lambda}\to 0} \frac{1}{\bar{\lambda}} G_{\bar{d}}\left(\frac{|\bar{x}-u|}{\bar{\lambda}}\right) = \tanh(\bar{d})\delta(\bar{x}-u)$ with the Dirac delta distribution δ and Eq. (2.75a) result in the spatial phase variation

$$\sin(\varphi) = \int_{-\infty}^{\infty} \tanh(\bar{d})\delta(\bar{x}-u) \frac{d^2\varphi}{du^2} du = \tanh(\bar{d}) \frac{d^2\varphi}{d\bar{x}^2}. \tag{2.76}$$

This result is a local sine-Gordon equation with the effective Josephson length

$$\lambda_J^{eff} = \lambda_J^b \sqrt{\tanh\left(\frac{d}{\lambda_L}\right)}. \tag{2.77}$$

The equal result is obtained from the Josephson length's definition in Eq. (2.73) by inserting an effective magnetic thickness in junctions with thin-film electrodes, as derived earlier by Weihnacht [Wei69]

$$\Lambda^{eff} = \frac{\lambda_{L1}}{\tanh\left(\frac{d_1}{\lambda_{L1}}\right)} + \frac{\lambda_{L2}}{\tanh\left(\frac{d_2}{\lambda_{L2}}\right)} + d_I \tag{2.78}$$

with identical bottom and top electrode and negligible tunnel barrier thickness d_I.

The Josephson vortex solution (2.74) remains valid for Eq. (2.76), though the characteristic length is λ_J^{eff} instead of λ_J^b [AP95]. It should be noted, that this solution still features a phase winding (fluxoid) quantization in 2π, which, however, does not necessitate quantized magnetic flux anymore. As the major part of the vortex's total 2π phase winding drops at the dominating kinetic inductance that does not couple to magnetic fields and implies incomplete screening, the magnetic flux possessed by a Josephson vortex Φ is thus significantly smaller than the magnetic flux quantum Φ_0. Therefore, this kind of vortex should be more correctly denoted as "fluxoid" instead of "fluxon" [Wil+22].

Magnetic diffraction pattern of the critical current

The occurrence of Josephson supercurrents is an interference effect of two macroscopic wave functions that may vary both in time and space. As already mentioned in Sec. 2.4.1 and what is known as AC Josephson effect, applied electric fields (DC voltages) make for a temporally rolling phase difference, according to the second Josephson equation (2.60)). This yields (oscillating) supercurrents that are periodic in time. Applied magnetic fields, however, cause a spatially rolling gauge-invariant phase difference, as described by Eq. (2.71). Therefore, if the self-field is small compared to the external field, which is the case for short junctions with $\ell < \lambda_\mathrm{J}$, the Josephson current across a rectangular junction of length ℓ and width w results in

$$I_\mathrm{J}(H) = \int_{-w/2}^{w/2} \mathrm{d}y \int_{-\ell/2}^{\ell/2} \mathrm{d}x j_\mathrm{c} \sin\left(\frac{2\pi\mu_0 H\Lambda}{\Phi_0}x + \varphi_0\right) = I_\mathrm{c}(0)\frac{\sin\left(\frac{\pi\Phi}{\Phi_0}\right)}{\frac{\pi\Phi}{\Phi_0}}\sin(\varphi_0). \tag{2.79a}$$

As can be seen in Eq. (2.79a), the Josephson current vs. magnetic field distribution generally equals the Fourier transform of the junction's cross-section [Tin04]. In particular, the magnetic diffraction pattern of a rectangular junction cross-section gives rise to a sinc-shaped pattern function. Maximizing the current with respect to φ_0, results in the critical current [Jos64]

$$I_\mathrm{c} = I_\mathrm{c}(0)\left|\frac{\sin\left(\frac{\pi\Phi}{\Phi_0}\right)}{\frac{\pi\Phi}{\Phi_0}}\right|, \tag{2.79b}$$

which resembles the Fraunhofer pattern in optical single-slit experiments, where two wave functions also interfere with varying path differences. In distributed Josephson junctions, this path difference originates in the magnetic field dependence of the gauge-invariant phase difference and modifies the interference amplitude of the two superconducting wave functions. Figure 2.13 shows both simulated and experimentally acquired magnetic diffraction pattern of short junction's critical currents.

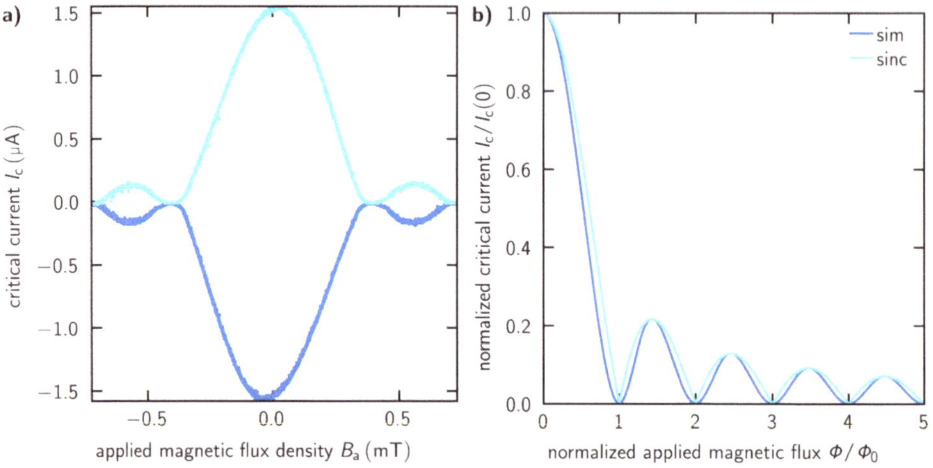

Figure 2.13: Fraunhofer-like magnetic diffraction pattern of a short junction's critical current. a) Measured diffraction pattern of a 80 μm × 5 μm junction in inline geometry with electrodes made from high-kinetic inductance AlO_x and pure aluminum. The slight asymmetry arises because of the inhomogeneous bias current distribution in the inline bias scheme in combination with different inductances in the bottom and top electrode [Sch70; BJV75; Mon+13]. These experimental data resemble **b)** the analytically calculated sinc-shaped diffraction Fraunhofer pattern, which, however, features dissonantly sharp turns at integer multiples of Φ_0. More advanced simulations [PRS91], assuming a normalized junction length $\ell/\lambda_J = 0.3$, cover this detail splendidly.

The simulated data are generated with the approach of Pagano *et al.* [PRS91] that is based on evaluating the phase space configuration $\varphi_{\bar{x}}(\varphi)$ of the perturbed Ferrell-Prange equation. In doing so, stable points can be found at a given applied field and for different fluxon branches, which yield extremal junction length depending on the applied bias current. This dependency is numerically inverted to find a critical current I_c at a fixed junction length ℓ.

In case of the above analytical derivation, the self-field, generated by the induced screening currents is completely neglected, such that the magnetic field inside the tunnel barrier equals the applied magnetic flux density $\mathbf{B} = \mu_0 \mathbf{H}$. This assumption is valid as long as the applied field penetrates the entire junction, which is the case, while the junction length ℓ exceeds the twice Josephson length λ_J, the so-called "short junction" limit $\ell \lesssim 2\lambda_J$[14] [OS67].

In the opposite regime of long junctions $\ell \gg \lambda_J$, however, self-fields become significant since screening currents repel magnetic fields from the junction's interior, in the same way as described by the Meissner-Ochsenfeld effect of bulk materials

[14] The factor 2 arises since the field penetrates the junction from both sides, as illustrated in Fig. 2.16 a).

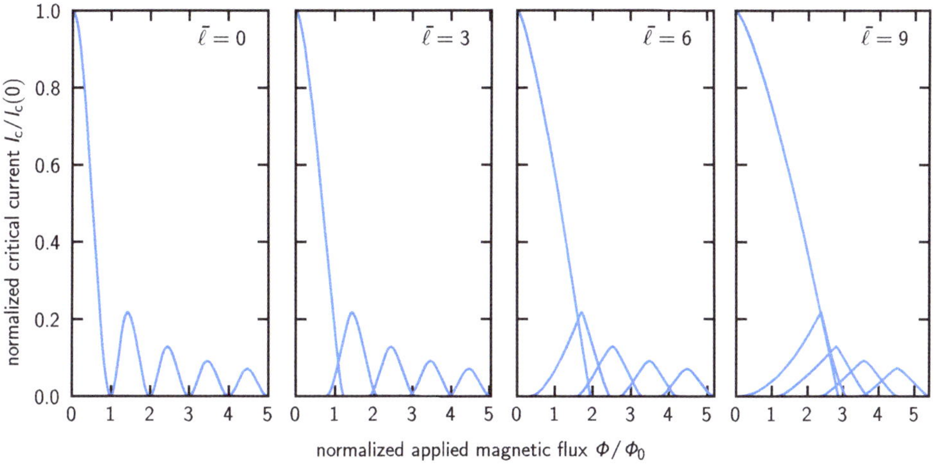

Figure 2.14: Magnetic diffraction pattern of the critical current for different junction lengths. The simulations follow the approach of Pagano *et al.* [PRS91] and assume a homogeneous bias current distribution, such as is the case for the overlap geometry. The initially rounded lobes of the Fraunhofer diffraction pattern of short junctions ($\bar{\ell} = 0$) become more and more linear with increasing junction lengths. Thereby, the main maximum widens and eventually covers the former adjacent higher-order side maxima. The different lobes correspond to the number of inherent flux quanta, while it should be stressed that junctions enter the state that may carry the maximal possible current to minimize their free energy, which is observed in experiments, too.

(see Sec. 2.1) Similar to type-II superconductors, in Josephson junctions, screening currents increase with the applied magnetic field to maintain perfect diamagnetism, which collapses above a first critical field H_{c1}. There, entering of a magnetic flux quantum in the junction is energetically favorable, and since screening currents only need to repel parts of the applied flux, they decrease. As a consequence, the junction can carry larger bias currents, and a side maximum occurs in the magnetic diffraction pattern. Increasing the magnetic field further lets the same effect start all over again, and each lobe corresponds to the number of inherent magnetic flux quanta, starting from zero in the case of the main maxima [OS67]. The Meissner-like linear increase in screening currents manifests in long junctions' magnetic diffraction pattern of the critical current as linear slopes of the individual lobes. As can be seen in Fig. 2.14, there is no sharp transition from short to long junctions with any hard criterion, but typically junctions are denoted as one-dimensional "long junction" for normalized junction lengths $\bar{\ell} = \ell/\lambda_J \gtrsim 7$, whereas short widths $w \ll \lambda_J$ do not allow for spatial phase variations along the y-direction [PRS91].

Aside from their normalized length, long Josephson junctions' exact magnetic diffraction pattern is defined by the junction geometry as this goes along with different current distribution [OS67; Sch70; BJV75; BP82]. There are three widespread main types of linear electrode configurations, whose schematic and length-dependent maximal critical current are shown in Fig. 2.15:

- overlap geometry: In this configuration, as shown in Fig. 2.15 a) (i), bias currents distribute homogeneously across the whole length, such that the junction's maximum Josephson current increases proportionally to the total junction area as

$$\frac{I_J}{2w\lambda_J j_c} = \frac{\ell}{2\lambda_J}. \tag{2.80a}$$

- inline geometry: In the symmetric inline arrangement, as depicted in Fig. 2.15 a) (ii), bias currents are injected and ejected close to the two edges over a length of λ_J, which is why the normalized maximum Josephson current holds

$$\frac{I_J}{2w\lambda_J j_c} = \frac{2\sinh\left(\frac{\ell}{2\lambda_J}\right)}{\cosh\left(\frac{\ell}{2\lambda_J}\right)+1}, \tag{2.80b}$$

  and thus converges towards 2 for long junctions. An asymmetric inline geometry, as illustrated in Fig. 2.15 a) (iii) is also possible, where only the one edge close to the bias leads is stressed, and thus the normalized maximum Josephson current is given by

$$\frac{I_J}{2w\lambda_J j_c} = \tanh\left(\frac{\ell}{2\lambda_J}\right), \tag{2.80c}$$

  saturating at 1 for long junctions.

- crossover geometry: This configuration, as schematically presented in Fig. 2.15 a) (iv), is a particular combination of inline and overlap configuration, where bias currents are injected along the short side (as in the case of inline geometry) and ejected along the long side (like in overlap junctions). If the Josephson inductance dominates the low lead inductance, the normalized maximum Josephson current is $1 \leq \lim_{\ell\to\infty} \frac{I_J}{2w\lambda_J j_c} \leq 2$, depending on the fraction of bias current flowing in each edge. In the particular case of high- and low-inductive leads in the y- and x-direction, this yields a homogeneous current distribution, which is why it is referred to as quasi-overlap geometry [Sar+91].

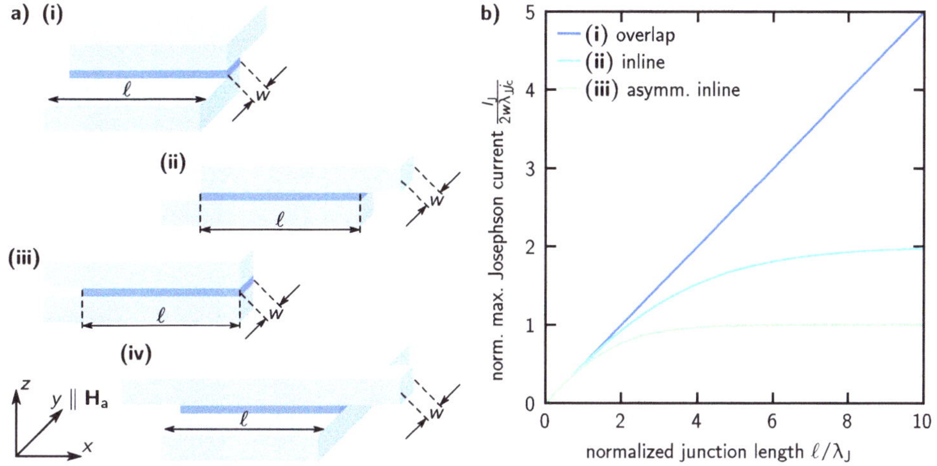

Figure 2.15: Schematic layout and maximum Josephson current for different junction geometries. **a)** Popular one-dimensional linear long junction geometries are referred to as **(i)** symmetric inline, **(ii)** asymmetric inline, **(iii)** overlap, and **(iv)** crossover. **b)** These configurations differ in their current distribution, for what reason, their maximum Josephson current varies significantly. Merely the overlap geometry provides a uniform current distribution, and thus its maximal Josephson current does not saturate in the long junction limit, as otherwise. This figure is adapted from Ref. [BJV75].

2.4.3 The sine-Gordon model of long junctions

To analyze phase dynamics, covering both spatial and temporal phase variations $\varphi(x,t)$, the junction capacitance needs to be considered similarly to the RCSJ model, but for distributed junctions with inductive electrodes [11]. Within this scope, one-dimensional junctions can be modeled by $N \to \infty$ infinitesimally small lumped elements of resistively and capacitively shunted junctions in the z-direction, which are extended along the x-axis, and thus connected via inductive and resistive leads. The finite resistance in parallel to the electrode's inductance is due to surface losses. As shown in Appendix A.1, the inductances of both electrodes can be mapped to a single effective inductance, resulting in the equivalent circuit illustrated in Fig. 2.16 b). Using common circuit quantization techniques [VD17] with N node fluxes Φ_j, the system holds the Lagrangian

$$\mathcal{L} = \sum_{j=0}^{N-1} \frac{c_J w \Delta x}{2} \left(\partial_t \Phi_j\right)^2 - \sum_{j=1}^{N-1} \frac{\left(\Phi_j - \Phi_{j-1} - \Phi_j^{\text{ext}} + \Phi_{j-1}^{\text{ext}}\right)^2}{2 L_0^{\square} \frac{\Delta x}{w}} \\ - \sum_{j=0}^{N-1} \frac{\Phi_0 j_c w \Delta x}{2\pi} \left(1 - \cos\left(\frac{2\pi}{\Phi_0} \Phi_j\right)\right) + \sum_{j=0}^{N-1} j_b w \Delta x \Phi_j \tag{2.81a}$$

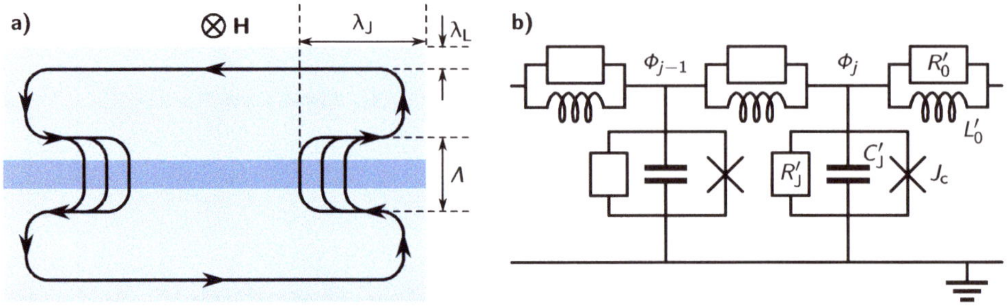

Figure 2.16: Schematic screening currents and equivalent circuit of long Joséphson junctions. a) According to the Meissner-Ochsenfeld effect, screening currents (black solid lines) are induced by an applied magnetic field, which penetrate the bulk electrodes on the scale of the London penetration depth λ_L. By comparison, superconductivity is suppressed in the junction region, for what reason the penetration depth there, the Josephson length λ_J, exceeds the λ_L significantly. Screening currents penetrate vertically on the scale of the magnetic thickness Λ that composes of the tunnel barrier's thickness d_I and the λ_L of each electrode. **b)** The Long junction's equivalent circuit is represented by lumped elements of resistively $\left(R_J'\right)$ and capacitively $\left(C_J'\right)$ shunted junctions (J_c) in the z-direction, which are extended along the x-axis, and thus connected via inductive (L_0') and resistive (R_0') leads.

$$\mathcal{L} = w \int_0^{\ell} \left[\frac{c_J}{2} \left(\partial_t \Phi\right)^2 - \frac{\left(\partial_x \Phi - A\right)^2}{2L_0^{\square}} - \frac{\Phi_0 j_c}{2\pi}\left(1 - \cos\left(\frac{2\pi}{\Phi_0}\Phi\right)\right) + j_b \Phi \right] \mathrm{d}x. \tag{2.81b}$$

Here, $c_J = C_J^{\mathrm{tot}}/w\ell$ denotes the specific junction capacitance, $L_0^{\square} = L_0^{\mathrm{tot}} w/\ell$ the electrode's sheet inductance that sums up both the geometric and kinetic parts, $j_c = I_c^{\mathrm{tot}}/w\ell$, and $j_b = I_b^{\mathrm{tot}}/w\ell$ describe the critical and bias current density, respectively. The quoted bias current is commonly driven by external current sources, but magnetic field induced currents $j_z = \frac{\mathrm{d}H_y(x)}{\mathrm{d}x}$ can be treated analogously.

During the transition to the continuous limit of distributed long junctions ($N \to \infty$, $\Delta x \to 0$), the sum is replaced by an integral $\sum_{j=0}^{N-1} \ldots \Delta x \to \int_0^{\ell} \ldots \mathrm{d}x$, the difference quotient becomes a derivative $\lim_{\Delta x \to 0} \frac{\Phi_j - \Phi_{j-1}}{\Delta x} = \frac{\mathrm{d}\Phi(x)}{\mathrm{d}x}$, and the external flux is transferred to the electromagnetic vector potential $\lim_{\Delta x \to 0} \frac{\Phi_j^{\mathrm{ext}} - \Phi_{j-1}^{\mathrm{ext}}}{\Delta x} = \frac{\mathrm{d}\Phi^{\mathrm{ext}}(x)}{\mathrm{d}x} = A(x)$. In order to describe the junction's phase dynamics, node fluxes are translated to node phases via the phase-flux relation $\Phi_j = \frac{\Phi_0}{2\pi}\varphi_j$.

With the assumption of static external magnetic flux $\partial_t \Phi^{\mathrm{ext}} = 0$, the phase difference can be redefined, shifted by electromagnetic vector potential A, so that $\varphi \to \varphi + \frac{2\pi}{\Phi_0}\int A \mathrm{d}z = \varphi - \varphi^{\mathrm{ext}}$ with the externally applied phase φ^{ext}. Finally, the long junction's phase dynamics are described by the sine-Gordon Lagrangian in normalized units

$$\bar{\mathcal{L}} = \int_{0}^{\bar{\ell}} \left[\frac{1}{2} (\partial_{\bar{t}}\varphi)^2 - \frac{1}{2} (\partial_{\bar{x}}\varphi)^2 - (1 - \cos(\varphi - \varphi^{\text{ext}})) + \gamma\varphi \right] \mathrm{d}\bar{x}. \tag{2.81c}$$

Here, the Josephson plasma frequency ω_p, the Josephson penetration depth λ_J, and $\mathcal{E}_\text{J}$ are the characteristic inverse time, length, and energy scales in the long Josephson junction, such that these quantities are normalized to

$$\bar{t} = \omega_\text{p} t, \qquad \omega_\text{p} = \sqrt{\frac{2\pi j_\text{c}}{\Phi_0 c_\text{J}}}, \tag{2.82a}$$

$$\bar{x} = x/\lambda_\text{J}, \qquad \lambda_\text{J} = \sqrt{\frac{\Phi_0}{2\pi j_\text{c} L^\square}}, \tag{2.82b}$$

$$\bar{\mathcal{L}} = \mathcal{L}/\mathcal{E}_\text{J}, \qquad \mathcal{E}_\text{J} = \frac{\Phi_0 j_\text{c} \lambda_\text{J} w}{2\pi}, \tag{2.82c}$$

$$\gamma = j_\text{b}/j_\text{c}. \tag{2.82d}$$

Applying the Euler-Lagrange equation [BP82; Leg87; CF91] yields the equation of motion [Jos65]

$$0 = \frac{\mathrm{d}}{\mathrm{d}\bar{t}} \frac{\partial \bar{\mathcal{L}}}{\partial_{\bar{t}}\varphi} + \frac{\mathrm{d}}{\mathrm{d}\bar{x}} \frac{\partial \bar{\mathcal{L}}}{\partial_{\bar{x}}\varphi} - \frac{\partial \bar{\mathcal{L}}}{\partial \varphi}, \tag{2.83a}$$

$$\gamma = \partial_{\bar{t}\bar{t}}\varphi - \partial_{\bar{x}\bar{x}}\varphi + \sin\varphi. \tag{2.83b}$$

This so-called sine-Gordon equation is a nonlinear wave equation, which describes dynamic properties in a Josephson transmission line. The characteristic propagation velocity inside this Josephson medium is denoted as Swihart velocity $\bar{c} = \omega_\text{p}\lambda_\text{J} = \frac{1}{\sqrt{L^\square c_\text{J}}}$ and is typically in the order of $\sim 10^{-2}\, c_0$. The corresponding junction impedance $Z_\text{J} = \frac{1}{w}\sqrt{\frac{L^\square}{c_\text{J}}} \propto \bar{c}^{-1}$ is consequently in the order of a few ohms.

For the purpose of Lagrangian formalism, dissipative perturbations that are unpreventable in real long junctions are usually included by a Rayleigh dissipation functional $\mathcal{R} = \int \frac{1}{2R} (\partial_t \Phi)^2 \,\mathrm{d}x$ [JV09; EGK15]. Introducing both the ohmic quasiparticle tunneling resistance $R_\text{J} = \rho_\text{J} d_\text{I}/w\Delta x = r_\text{J}/w\Delta x$ and the lateral electrode resistance due to surface losses in the superconductor $R_0 = \rho_0 \Delta x/wd = R_0^\square \Delta x/w$, as depicted in Fig. 2.16 b), results in

$$\mathcal{R} = \sum_{j=0}^{N-1} \frac{w\Delta x}{2\rho_\text{J} d_\text{I}} (\partial_t \Phi_j)^2 + \sum_{j=1}^{N-1} \frac{w}{2R_0^\square \Delta x} (\partial_t \Phi_j - \partial_t \Phi_{j-1})^2 \tag{2.84a}$$

$$= w \int_0^\ell \left(\frac{1}{2r_\text{J}} (\partial_t \Phi)^2 + \frac{1}{2R_0^\square} (\partial_t \partial_x \Phi)^2 \right) \mathrm{d}x. \tag{2.84b}$$

With the typical normalization of time, length, and energy, given in Eq. (2.82), the dimensionless Rayleigh dissipation functional results in [MS78]

$$\frac{\bar{\mathcal{R}}}{\omega_\mathrm{p}} = \frac{1}{2}\int_0^{\bar{\ell}} \left(\alpha\left(\partial_{\bar{t}}\varphi\right)^2 + \beta\left(\partial_{\bar{x}\bar{t}}\varphi\right)^2\right) \mathrm{d}\bar{x} \tag{2.84c}$$

with the dimensionless damping coefficients covering quasiparticle tunneling across the tunnel barrier α and electrode's surface losses β

$$\alpha = \sqrt{\frac{\Phi_0}{2\pi j_\mathrm{c} r_\mathrm{J}^2 c_\mathrm{J}}} = \beta_\mathrm{c}^{-1/2}, \tag{2.85a}$$

$$\beta = \sqrt{\frac{2\pi j_\mathrm{c}\left(L_0^\square\right)^2}{\Phi_0 c_\mathrm{J}\left(R_0^\square\right)^2}}. \tag{2.85b}$$

The junction's Hamiltonian, governing its total energy, can be obtained by a classical Legendre transformation and is commonly split in a sine-Gordon summand $\bar{\mathcal{H}}^\mathrm{SG}$ and a perturbation summand $\bar{\mathcal{H}}^\mathrm{P}$ including bias and dissipation

$$\bar{\mathcal{H}} = \bar{\mathcal{H}}^\mathrm{SG} + \bar{\mathcal{H}}^\mathrm{P}, \tag{2.86a}$$

$$\bar{\mathcal{H}}^\mathrm{SG} = \int_0^{\bar{\ell}} \left[\frac{1}{2}\left(\partial_{\bar{t}}\varphi\right)^2 + \frac{1}{2}\left(\partial_{\bar{x}}\varphi\right)^2 + \left(1-\cos(\varphi-\varphi_\mathrm{ext})\right)\right] \mathrm{d}\bar{x}, \tag{2.86b}$$

$$\frac{\mathrm{d}\bar{\mathcal{H}}^\mathrm{P}}{\mathrm{d}t} = \int_0^{\bar{\ell}} \left[\gamma\varphi_{\bar{t}} - \alpha\varphi_{\bar{t}}^2 - \beta\varphi_{\bar{x}\bar{t}}^2\right] \mathrm{d}\bar{x}. \tag{2.86c}$$

The Euler-Lagrange equation including dissipation by means of the Rayleigh functional reads [GPS14; Nol14]

$$\frac{\mathrm{d}}{\mathrm{d}\bar{t}}\left(\frac{\partial\bar{\mathcal{L}}}{\partial\left(\partial_{\bar{t}}\varphi\right)}\right) + \frac{\mathrm{d}}{\mathrm{d}\bar{x}}\left(\frac{\partial\bar{\mathcal{L}}}{\partial\left(\partial_{\bar{x}}\varphi\right)}\right) - \frac{\partial\bar{\mathcal{L}}}{\partial\varphi} = \frac{1}{\omega_\mathrm{p}}\frac{\mathrm{d}}{\mathrm{d}\bar{x}}\left(\frac{\partial\bar{\mathcal{R}}}{\partial\left(\partial_{\bar{x}\bar{t}}\varphi\right)}\right) - \frac{1}{\omega_\mathrm{p}}\frac{\partial\bar{\mathcal{R}}}{\partial\left(\partial_{\bar{t}}\varphi\right)} \tag{2.87}$$

and results in the perturbed sine-Gordon equation[MS78]

$$\partial_{\bar{t}\bar{t}}\varphi - \partial_{\bar{x}\bar{x}}\varphi + \sin\varphi = \gamma - \alpha\partial_{\bar{t}}\varphi + \beta\partial_{\bar{x}\bar{x}\bar{t}}\varphi. \tag{2.88}$$

In order to describe the phase dynamics of long junctions with finite length properly, magnetic fields, influencing the spatial phase dependence $\varphi(\bar{x})$, need to be analyzed. The part of externally applied fields in the y-direction H_a always affects $\varphi(\bar{x})$ according to Eq. (2.71), whereas the part H_b, induced by a bias current I_b, strongly depends on the junction's geometry [MS78; LPS83; Ols+86].

In case of the overlap configuration (see Fig. 2.15 a) (i)) the bias current both in the top and bottom electrode is applied in the y-direction, hence, the induced H_b is bound to the xz-plane. Since, it is perpendicular to the junction's area, the z-component does not influence the gauge-invariant phase difference at all. Due to the small junction width $w \ll \lambda_J$, the magnetic flux through the yz-plane, generated by the x-component of the induced magnetic field, is negligible small for phase variation on the scale of λ_J. That is, the bias current induced magnetic fields do not affect the phase junction's dynamics that is fully governed by the externally applied field according to Eq. (2.71), such that the boundary conditions of overlap junctions hold

$$\left(\varphi_{\bar{x}} + \beta\varphi_{\bar{x}\bar{t}}\right)\Big|_{\bar{x}=0} = \left(\varphi_{\bar{x}} + \beta\varphi_{\bar{x}\bar{t}}\right)\Big|_{\bar{x}=\bar{\ell}} = \frac{2\pi\mu_0\Lambda\lambda_J}{\Phi_0}H := \eta. \tag{2.89}$$

The situation changes for the inline geometry (see Fig. 2.15 a) (ii)), where the bias currents flow in parallel to the junction's long side along x and its induced magnetic fields are in the yz-plane. As for overlap junctions, the z-component does not affect $\varphi(\bar{x})$, but the y-component contributes in the same manner as the H_{a_y} does. Inside the junction ($0 < \bar{x} < \bar{\ell}$), H_b of the bottom and top electrode cancel to zero, but at one junction end, the H_b above the bottom electrode and at the other end H_b below the top electrode contribute with different signs. Therefore, the boundary conditions of symmetric inline junctions are given by

$$\left(\varphi_{\bar{x}} + \beta\varphi_{\bar{x}\bar{t}}\right)\Big|_{\bar{x}=0} = \frac{2\pi\mu_0\Lambda\lambda_J}{\Phi_0}H + \frac{I_b}{2j_c\lambda_J w} := \eta + \gamma\frac{\bar{\ell}}{2}, \tag{2.90a}$$

$$\left(\varphi_{\bar{x}} + \beta\varphi_{\bar{x}\bar{t}}\right)\Big|_{\bar{x}=\bar{\ell}} = \frac{2\pi\mu_0\Lambda\lambda_J}{\Phi_0}H - \frac{I_b}{2j_c\lambda_J w} := \eta - \gamma\frac{\bar{\ell}}{2}, \tag{2.90b}$$

or vice versa, depending on whether both dues sum up constructively or destructively.

Small-amplitude plasma wave solution

The generally nonlinear sine-Gordon equation (2.83b) of an infinite length Josephson junction can be linearized for small amplitudes $\varphi \ll 1$ and becomes a Klein-Gordon equation

$$\varphi_{\bar{x}\bar{x}} - \varphi_{\bar{t}\bar{t}} = \varphi, \tag{2.91}$$

whose solution are plane waves with normalized wave number $\bar{k} = k \cdot \lambda_J$, frequency $\bar{\omega} = \omega/\omega_p$, and small amplitude $\varphi_0 \ll 1$

$$\varphi_p(\bar{x}, \bar{t}) = \varphi_0 e^{i(\bar{k}\bar{x} - \bar{\omega}\bar{t})} + 2\pi n. \tag{2.92}$$

The propagation of these small amplitude excitations in space and time is illustrated in Fig. 2.17 a). Inserting this plane wave solution of Eq. (2.92) in the Klein-Gordon equation (2.91) yields the dispersion relation $\bar{\omega} = \sqrt{1 + \bar{k}^2}$ (see Fig. 2.17 b)), which resembles plasma oscillations since it features a gap in the spectrum at a small $\bar{k}$ of 1 (of ω_p in SI units). This is the reason why such a linear excitation is commonly

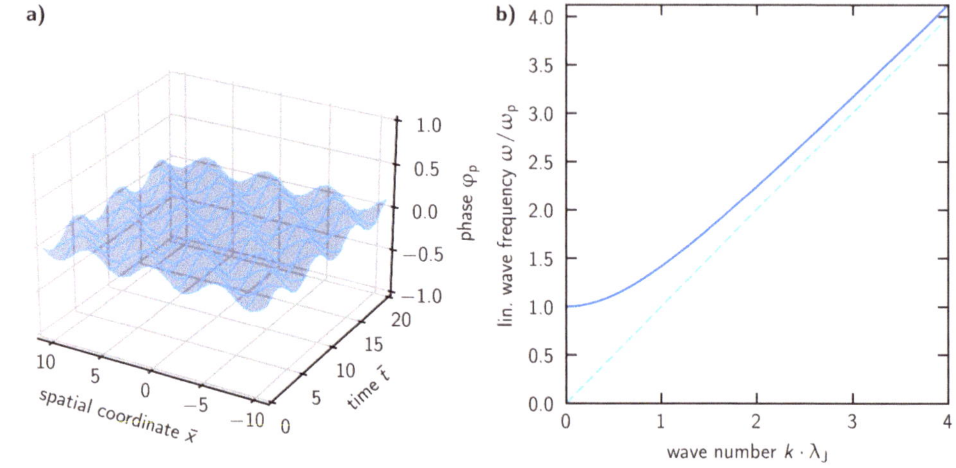

Figure 2.17: Small amplitude plane wave solution of the linearized sine-Gordon equation. a) Plane waves, solving the linearized sine-Gordon equation for small amplitudes $\varphi_0 = 0.1 \ll 1$, propagate both in space with the wave number $\bar{k} = 1$ and in time with the frequency $\bar{\omega} = 1$. **b)** Their dispersion relation exhibits a gap for small wave numbers $\bar{k}$ and approaches a linear course for $\bar{k} \gg 1$.

called "Josephson plasma oscillation" or simply "plasmon" [OD91] and the former introduced oscillation frequency in Josephson junctions is referred to as "Josephson plasma frequency".

Josephson vortex solution

The solution of infinite length junctions, which attracted most attention in the last decades, is the Josephson vortex solution. Neglecting perturbations on the right side of Eq. 2.88, this sine-Gordon equation is solved by

$$\varphi_F(\bar{x}, \bar{t}) = 4 \arctan\left(\exp\left(\pm\frac{\bar{x} - \bar{v}\bar{t} - \bar{x}_0}{\sqrt{1 - \bar{v}^2}}\right)\right) + 2\pi n, \tag{2.93}$$

which is nothing but the dynamic version of Josephson vortices of Eq. (2.74) discussed in Sec. 2.4.2, however, traveling along the junction with velocity $\bar{v} = v/\bar{c}$. This means that the sign of the exponent of Eq. (2.93) determines the direction of propagation, manifesting as vortex or antivortex with opposite ring current.

These Josephson vortices behave absolutely as solitons [Ust98], which are self-reinforcing wave packages with permanent shape due to balanced perturbations while propagating a medium. This requires nonlinear effects that cancel dispersive effects in a manner that the wave package's faster frequency components are converted into slower ones and vise versa, resulting in the dynamic equilibrium of the soliton. The propagation velocity is limited to the Swihart velocity $\bar{c}$, the

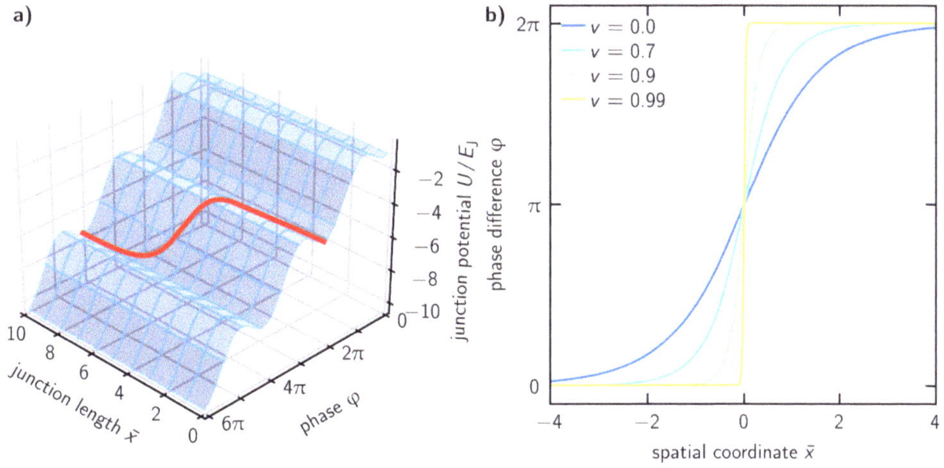

Figure 2.18: Josephson vortex solution in long Josephson junctions. a) The 2π phase kink solution (red line) spans over two neighboring valleys in the tilted washboard potential of a spatially extended long junction. **b)** At relativistic velocities, the Josephson vortex reveals Lorentz contraction such that the 2π-kink becomes sharper and sharper (from dark to bright).

speed of light inside the Josephson medium. As can be seen from the Lorentz factor $\left(1-\bar{v}^2\right)^{-1/2}$ in Eq. (2.93), these solitons are invariant with respect to the Lorentz transformation

$$\bar{x} \to \tilde{x} = \frac{\bar{x} - \bar{v}\bar{t}}{\sqrt{1-\bar{v}^2}}, \tag{2.94a}$$

$$\bar{t} \to \tilde{t} = \frac{\bar{t} - \bar{x}/\bar{v}}{\sqrt{1-\bar{v}^2}}, \tag{2.94b}$$

since both the d'Alembert operator and $\sin(\varphi)$ in the unperturbed sine-Gordon equation (2.83b) are Lorentz invariant. As a consequence, the solitons behave like relativistic particles, exhibiting Lorentz contraction (see Fig. 2.18 b)) for velocities approaching the Swihart velocity. Furthermore, Josephson vortices' total energy are given by

$$\bar{E}_{\mathrm{F}} = \bar{\mathcal{H}}(\varphi_{\mathrm{F}}) = \frac{8}{\sqrt{1-\bar{v}^2}}, \tag{2.95}$$

which features a relativistic mass increase and the normalized rest mass $\bar{m}_{\mathrm{F}} = 8$ [MS78]. The soliton propagation can be driven up to relativistic velocities by external forces, such as a current bias that acts via a Lorentz-Magnus force on the vortex's participating charge carriers. This acceleration is accompanied by dissipation that yields an equilibrium velocity $\bar{v}_\infty$, which can be determined by equating the time derivatives of the sine-Gordon and perturbation Hamiltonians

$$\left.\frac{\mathrm{d}\bar{\mathcal{H}}^{\mathrm{SG}}}{\mathrm{d}\bar{t}}\right|_{\varphi=\varphi_\mathrm{F}} = \frac{8\bar{v}}{\left(1-\bar{v}^2\right)^{3/2}}\frac{\mathrm{d}\bar{v}}{\mathrm{d}\bar{t}}, \tag{2.96a}$$

$$\left.\frac{\mathrm{d}\bar{\mathcal{H}}^{\mathrm{P}}}{\mathrm{d}\bar{t}}\right|_{\varphi=\varphi_\mathrm{F}} = -8\alpha\frac{\bar{v}^2}{\left(1-\bar{v}^2\right)^{1/2}} - \frac{8}{3}\beta\frac{\bar{v}^2}{\left(1-\bar{v}^2\right)^{3/2}} - 2\pi\gamma, \tag{2.96b}$$

and results in the first-order ordinary differential equation for $\bar{v}(\bar{t})$ [MS78]

$$\frac{\mathrm{d}\bar{v}}{\mathrm{d}\bar{t}} = -\alpha\bar{v}\left(1-\bar{v}^2\right) - \frac{1}{3}\beta\bar{v} - \frac{\pi}{4}\gamma\left(1-\bar{v}^2\right)^{3/2}. \tag{2.97}$$

The situation, where the input power supplied by the external current source equals the dissipated power, is characterized by $\left.\frac{\mathrm{d}\bar{v}}{\mathrm{d}\bar{t}}\right|_{\bar{v}=\bar{v}_\infty} = 0$ with the power balance velocity $\bar{v}_\infty$ and yields

$$\gamma = \frac{4}{\pi}\frac{|\bar{v}_\infty|}{\sqrt{1-\bar{v}_\infty^2}}\left(\alpha + \frac{\beta}{3\left(1-\bar{v}_\infty^2\right)}\right). \tag{2.98}$$

2.4.4 Quantumness of long Josephson junctions

From a more quantum mechanical perspective, the long junction's Lagrangian is conveniently written in the form [HSB94; SBM97]

$$\mathcal{L} = \frac{\hbar\bar{c}}{\bar{\hbar}_\mathrm{J}}\int_0^\ell \frac{1}{2\bar{c}^2}\left(\partial_t\varphi\right)^2 - \frac{1}{2}\left(\partial_x\varphi - A\right)^2 - \frac{1}{\lambda_\mathrm{J}^2}\left(1-\cos(\varphi)\right)\mathrm{d}x. \tag{2.99}$$

The overall coefficient

$$\bar{\hbar}_\mathrm{J} = \frac{\hbar\omega_\mathrm{p}}{\mathcal{E}_\mathrm{J}} = \frac{4\pi^2\hbar}{\Phi_0^2}\sqrt{\frac{L^\square}{c_\mathrm{J}w^2}} \propto Z_\mathrm{J} \tag{2.100}$$

defines the system's energy scale with respect to its quantized eigenenergies. In doing so, $\bar{\hbar}_\mathrm{J}$ can be interpreted as a normalized Planck constant in the Josephson medium [KWU02] that is a measure for the quantumness of LJJs. It is zero in the classical limit, whereas the larger $\bar{\hbar}_\mathrm{J}$, the smaller the system's energy scale, and thus the more quantum the junction is. For a typical 1 µm wide $\mathrm{Al}/\mathrm{AlO}_x/\mathrm{Al}$-junction, this normalized Planck constant is about $\bar{\hbar}_\mathrm{J} \approx 2\times10^{-3}$ [KI96], which means that energy scale of inherent excitations such as Josephson vortices exceed the quantized eigenenergies by $1/\bar{\hbar}_\mathrm{J}$. In consequence, such a conventional Josephson medium and its excitations are quite classical according to the Ehrenfest theorem.

The quantum regime can be approached by increasing the junction impedance by either narrowing the sample, decreasing the junction's specific capacitance, or increasing the electrode's inductance. In previous works, the junction widths w were reduced down to $\sim$ 200 nm–300 nm, yielding normalized Planck constants in the range $\bar{\hbar}_J \approx 7 \times 10^{-3}$–$9 \times 10^{-3}$ [Wal01; Kem06]. The further narrowing is experimentally challenging since the state-of-the-art fabrication technology limits the feasible structure sizes. Moreover, the widths cannot be reduced arbitrarily to ensure well-defined phases in the superconducting electrodes by avoiding phase slips [Vos+21]. The electrode's geometric inductance $L_g^\square$ is mostly defined by the actual shape and dimensions and is difficult to increase further. The specific capacitance is bound to suitable tunnel barrier materials that also control the critical current density and quasiparticle dissipation so that these three parameters are mainly exhausted.

However, the advent of high-kinetic inductance materials, exceeding the geometric inductance by orders of magnitude, is a promising, additional parameter for $\bar{\hbar}_J$ that relaxes the requirement on the lithographic fabrication resolution. The approach, taken throughout this work, to increase the junction's quantumness is the replacement of the junction's bulk electrodes with thin films of very high-kinetic inductance superconductors.

3 Experimental methods

The replacement of bulk long junction electrodes with a high-kinetic inductance superconductor is a promising approach for increasing the junction's impedance and its concomitant quantumness (see Sec. 2.4.4). In the following, the experimental realization chosen for this work is reported. There, the choice of granular aluminum as a high-kinetic inductance superconductor is motivated by elaborating the main properties of this material. After that, details of fabrication techniques and the measurement framework are specified. The latter consists of a refrigerator to cool the sample to cryogenic temperatures with inherent low-noise DC and RF wiring that ensure an appropriate environment. Finally, measurement electronics for transport or microwave measurements outside the cryostat are sketched out.

3.1 Fabrication techniques for high-impedance long junctions

Since most superconducting parameters are material-specific, the junction's electrode material, providing high kinetic inductance, plays a crucial role in the experimental realization. As seen in Tab. 2.1, thin films of granular aluminum oxide (AlO_x) are promising candidates since their large kinetic inductance exceeds the geometric due by orders of magnitudes. This is mandatory to increase the junction's impedance substantially. The combination of the material's adjustable large kinetic inductances with good microwave coherence of circuits made from AlO_x [Rot+16; Grü+18; Grü+19; Sch+20], are the main arguments for the usage of granular aluminum oxide for high-quality and high-impedance Josephson junctions. This material further turns out to be very suitable for making long Josephson junctions because of the good compatibility with common AlO_x tunnel barriers that are most widely used for quantum junctions due to performance and the straightforward native oxidation. In addition, AlO_x is resilient to magnetic in-plane fields [CA68; Chu+81b; Bor+20], which, for instance, are required to inject Josephson vortices in long junctions. Moreover, the critical temperature $T_c \approx 1.6\,\mathrm{K}$–$2.0\,\mathrm{K}$ well exceeds typical dilution refrigerators' base temperatures of $\sim 10\,\mathrm{mK}$, avoiding thermal quasiparticle excitations.

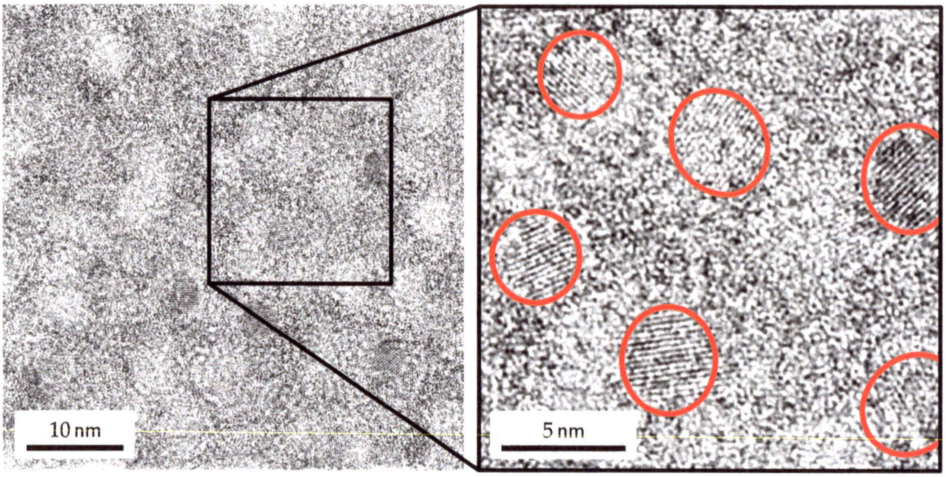

Figure 3.1: Transmission electron microscopy photograph of granular aluminum oxide. The highlighted areas with parallel stripes in the diffraction pattern are identified as crystalline pure aluminum grains with an approximated diameter of 3 nm–4 nm. These nanocrystals are randomly embedded in an insulating amorphous aluminum oxide matrix, such that this granular material can be regarded as disordered network of (super-) conducting islands separated by tunnel barriers.

3.1.1 Thin films of granular aluminum oxide

Initiated by the superconducting isotope effect and suggestions of electron-lattice interactions that are affecting the superconducting behavior, superconductors have been studied in the dirty limit for decades. In doing so, Buckel and Hilsch found, pure aluminum, deposited on a 4 K cold quartz substrate, to form a granular structure and to reveal a increased transition temperature up to $T_c = 2.59\,\mathrm{K}$ [BH54]. The grain size can be reduced, while increasing the critical temperature to $T_c \approx 3\,\mathrm{K}$ by depositing aluminum inside an oxygen atmosphere [ACC66].

Such granular aluminum oxide is used throughout this work and consists of pure crystalline aluminum grains embedded in insulating amorphous aluminum-oxide, as depicted in Fig. 3.1. Both the normal and the superconducting transport properties are dominated by these intrinsic nanoscopic tunnel barriers [Deu+73a]. In the superconducting phase, this granular structure can be treated as a network of nanoscopic Josephson junctions [Deu+73a] providing a kinetic-type Josephson inductance [Jos64], which is related to the normal state tunnel resistance R_n and the superconducting gap Δ by $L_k = \hbar R_n / \pi \Delta$ [MB58; GT57; Ann+10; Rot+16] (see Eq. (2.32)). The AlO_x's conductivity and inductance can range over five orders of magnitude [Deu+73b; Chu+81a] depending on the oxygen concentration in the nanoscopic tunnel barriers (TBs) [ZHB78], which is controlled by the oxygen partial pressure during the reactive sputtering process [Rot+16]. This enormous versatility enables us to use AlO_x in different regimes for depositing junction stacks with high kinetic inductance as bottom electrode, an insulating tunnel barrier, and a top electrode made from AlO_x and pure aluminum.

3.1.2 Trilayer sputter deposition

The junction trilayers are deposited on a c-plane sapphire substrate using a reactive DC magnetron sputter process in a homemade sputter tool. This device consists of an ultrahigh vacuum chamber to avoid contamination and a sputter gun formed by a ring-shaped anode close to the aluminum sputter target that serves as the cathode. Applying sufficiently high voltages $\sim 400\,\mathrm{V}$ enables a dielectric breakdown of the argon working gas, ignites an argon plasma that bombards the sputter target, and thus eject clusters of aluminum atoms. The sputtering rate is increased via a radial magnetic field that deflects charge carriers on an elongated spiral track, leading to an intensified plasma [DM08].

To grow AlO_x, oxygen is added as a reactive gas that oxidizes aluminum naturally and yields the desired granularity. The oxygen concentration, thus the normal sheet resistance $R_\mathrm{n}^\square$, and the concomitant kinetic inductance per unit square $L_\mathrm{k}^\square$, are defined by the oxygen partial pressure during the deposition [Rot+16]. Especially for sheet resistances beyond $\sim 1\,\mathrm{k\Omega}$, the resistivity is so sensitive to slight variations of the oxygen partial pressure that the reproducibility suffers from this exponential dependence.

To gain necessary control in this regime, both film thickness d and resistance $R_\mathrm{n}^\square$ are monitored during the film deposition.[1] The film thickness is measured using a quartz oscillator, whose resonant frequency shifts with the mass of deposited material. This quartz oscillator is placed close to the sample, and the frequency shift to thickness conversion is calibrated using a profilometer.

Conventional four-point probes of the film resistance are falsified by large currents $\sim 10\,\mathrm{mA}$ that flow from the plasma to the sample. To overcome this impact, large measurement signals, exceeding this plasma current by orders of magnitudes, cannot be used since they heat the film beyond its melting point. Such heating results in an inhomogeneous resistivity with undefined granularity and even bubbles on the film surface. Therefore, a counter voltage $\sim 200\,\mathrm{V}$ between the

[1] This *in situ* $R(d)$ measurement enables the estimation of the bottom electrode's sheet inductance, which would not be possible for the whole trilayer.

insulated sample holder and the grounded chamber is applied to decelerate the argon plasma, and thus avoid targeting the film and falsifying the measurement. The uncharged sputter material is unaffected by this electrical field, and the sputter rate remains the same.

This *in situ* $R(d)$ measurement enables us to fit the specific conductance $\sigma(d) \equiv \rho^{-1} = \left(R_{\mathrm{n}}^{\square} d\right)^{-1}$ and to estimate the final specific resistance ρ_{n0} already during the deposition [Wil+22]. The granular aluminum oxide's thickness-dependent specific conductance follows the model for fine-grained polycrystalline thin films by Mayadas *et al.* [MSJ69]

$$\frac{\sigma}{\sigma_0} \equiv \frac{\rho_{\mathrm{n0}}}{\rho_{\mathrm{n}}} = 3\left[\frac{1}{3} - \frac{\alpha}{2} + \alpha^2 - \alpha^3 \ln\left(1 + \frac{1}{\alpha}\right)\right]. \tag{3.1}$$

Equation (3.1) is the solution of a linearized Boltzmann equation including ordinary scattering mechanisms as in bulk materials and superimposed scattering at grain boundaries. There σ_0 is the intrinsic thickness-independent conductivity from the film interior and $\alpha := \frac{l_0}{d}\frac{r}{1-r}$ denotes the ratio between the background mean free path l_0 and the film thickness d, reduced by a scattering reflection coefficient r.

For the *in situ* fit, as shown in Fig. 3.2, an offset thickness has to be included, above which the model holds. The fit performed for trilayer A yields the resistivity $\rho_0 = (70.7 \pm 0.2)\,\mu\Omega\,\mathrm{cm}$ and $\frac{l_0 r}{1-r} = (5.04 \pm 0.17)\,\text{Å}$. The product of resistivity and mean free path is a material specific measure of its Fermi surface S_{F} and can be assumed to be $\rho_0 l_0 = 12\pi^3\hbar/e^2 S_{\mathrm{F}} = 1.6 \times 10^{-11}\,\Omega\,\mathrm{cm}^{-2}$ for granular aluminum oxide [CA68]. This yields the mean free path $l_0 = (22.62 \pm 0.07)\,\text{Å}$ in the order of a few nanometers and the substantial reflection coefficient $r = (18.2 \pm 0.5)\,\%$, which confirms diffusive transport in the granular material. The discrepancy between measurement and theoretical model for small film thicknesses originates in an inhomogeneous film thickness and the conductivity of the not perfectly repelled argon plasma that contributes especially for small film thicknesses, where the film is not entirely connected.

As can be seen in Fig. 3.2, this *in situ* $R(d)$ measurement is a powerful tool to achieve the aimed kinetic inductance value. On the one hand, the deposition can be stopped when reaching the desired sheet resistance, subjecting the exact film thickness to slight variations. On the other hand, if the fitted final resistivity deviates from the goal, the oxygen partial pressure can be readjusted, visible as knees in the $R(d)$ curves for trilayers B and C in Fig. 3.2. Consequently, the desired sheet resistance $R_{\square}^{\mathrm{n}}$ at the fixed film thickness can be achieved with a deviation of less than 10 % [Wil+22], which is great progress in controlling the sheet resistance of granular aluminum oxide.

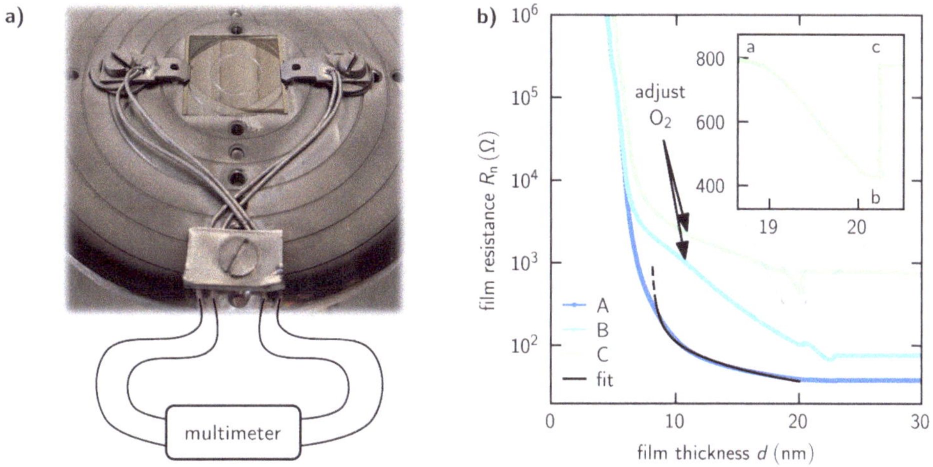

Figure 3.2: *In situ* resistance measurement during reactive sputter deposition of different trilayers. **a)** On two opposing sides of the substrate, there are previously deposited silver strips to ensure that the resistance of the squared film equals the sheet resistance, which is connected to a four-point measurement. **b)** The measured thickness-dependent normal resistance $R_n(d)$ is fitted to the model of fine-grained polycrystalline thin films by Mayadas *et al.* [MSJ69] in order to estimate the final resistance. If needed, the oxygen partial pressure is readjusted during the deposition to achieve the desired resistance, as is the case for trilayers B and C. The inset points out the tunnel barrier creation, where low-ohmic pure aluminum is deposited (a to b) and afterward statically oxidized (b to c). (This Figure is also published in Ref. [Wil+22].)

The trilayers' tunnel barriers are formed by depositing a thin film of pure aluminum, that is later oxidized statically. There, the critical current density is controlled by the product of oxidation pressure p_{ox} and time t_{ox} [KMM95]

$$j_c(p_{ox}, t_{ox}) = a_{ox} \left(p_{ox} \cdot t_{ox}\right)^{b_{ox}} . \tag{3.2}$$

Fitting results of the later transport measurements yield the constants $a_{ox} = (3.36 \pm 0.12) \times 10^4$ and $b_{ox} = -0.666 \pm 0.016$, when using the SI units ampere, meter, bar and second. For details, see Fig. B.2, given in Appendix B.2.

The barrier thickness can be estimated from $R(d)$ *in situ* measurement. The pure aluminum layer as a low-ohmic shunt first decreases the total sheet resistance (from a to b in the inset of Fig. 3.2) before coming back to the initial resistance at the end of the static oxidation of the whole aluminum layer (from b to c in the inset of Fig. 3.2). A complete oxidation is achieved for $\approx 1.2\,\text{nm}$–$1.5\,\text{nm}$ thick aluminum layers, but slightly depends on the barrier's transparency.

Finally, a top electrode is deposited similar to the bottom electrode but eventually with different oxygen concentrations. The characteristic parameters of the trilayers used in this work for long junction experiments are given in Tab. 3.1.

Table 3.1: Trilayer fabrication parameters. During the deposition of the bottom electrode, the sheet resistance $R_n^\square$ is measured by four-point probes, and the film thickness d_1 is measured with a quartz oscillator. To create the tunnel barrier, a layer of pure aluminum of thickness d_I is first sputtered and statically oxidized in an oxygen partial pressure p_{ox} for a time t_{ox}. The layer of pure aluminum is then deposited on the bottom electrode. If necessary, the top electrode of thickness d_2 is continued with another layer of thickness d_3.

trilayer	$R_n^\square$ (Ω)	p_{ox} (bar)	t_{ox} (min:s)	d_1 (nm)	d_I (nm)	d_2 (nm)	d_3 (nm)
A	38	21.0	20:00	20.4	2.6	20.6	20.0
B	75	0.51	4:00	20.4	2.1	21.1	20.2
C	778	5.1	4:00	18.6	1.5	20.4	20.1
D	2295	0.025	4:00	34.6	1.2	30.0	
E	947	0.286	4:00	46.6	1.2	30.0	

Long Josephson junctions are patterned from these stacks using e-beam and photolithography techniques in combination with inductively coupled plasma etching, anodic passivation and thermal evaporation of leads to the junction's top electrode. As the exact process differs for junctions read out in the DC or the RF scheme, this is discussed in detail in Sec. 4.2, Sec. 5.2, and Sec. 6.2.

3.2 Measurement setup

To study the physics of long Josephson junction devices, an appropriate framework of low-noise DC and RF connections as well as proper static magnetic fields are required. Therefore, the long junction samples are hosted in a gold-plated oxygen-free copper sample holder that provides two coaxial SMP jacks, which are soldered to $50\,\Omega$ matched microstrip transmission lines on microwave-suitable printed circuit boards (PCB). Aluminum wire bonds both connect these PCBs with the on-chip feedline and the DC microstructures with another PCB that supplies 24 DC lines from a micro D metal (MDM) connector. Moreover, a home-made superconducting niobium coil ($T_c \approx 9.2\,\mathrm{K}$) is placed around the sample holder to apply an in-plane magnetic field. To shield external magnetic fields, the whole box is surrounded by a Cryoperm shield.[2] The assembly used throughout this work is depicted in Fig. 3.3.

[2] Cryoperm is low-temperature suitable permalloy made from a soft magnetic nickel-copper-molybdenum-iron alloy. Due to its high permeability $\mu_r \approx 70\,000$ at $4\,\mathrm{K}$ and the entailing low reluctance, magnetic flux prefers being inside the shield instead of in the vacuum at its inner.

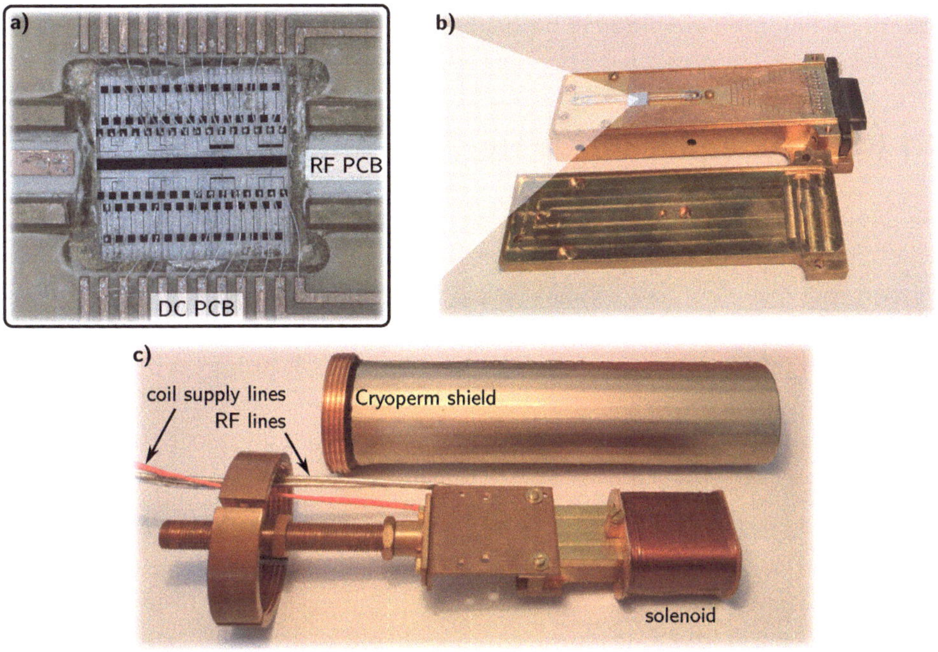

Figure 3.3: Sample holder with DC and RF connectors, surrounded by a magnetic coil and a permalloy shield. **a)** The sample chip, containing long junction experiments, is embedded in a sample holder, where aluminum wire bonds contact both junction's leads and microwave feedline to appropriate PCBs installed in the sample holder. **b)** The DC lines are connected to an MDM connector, whereas the RF-suitable PCBs are soldered to SMP jacks (accessible from the sample holder's bottom side, and thus not visible on this photograph). The hosting sample holder and its lid are made from copper and gold-plated to provide a good thermal connection to the cryostat. **c)** This closed box is surrounded by a superconducting coil to apply well-defined magnetic fields and mounted deep inside a Cryoperm shield, which screens magnetic fields from the sample.

Naturally, Josephson junction devices, made from aluminum (oxide) trilayers, necessitate cooling down to millikelvin temperatures. First superconducting phenomena, such as zero DC resistance, ideal diamagnetism and even coherent Josephson tunneling, already occur just below T_c. However, avoiding thermally populated dissipative quasiparticles and thermally activated tunnel processes that would destroy quantum coherence require way colder temperatures.

^{3}He/^{4}He dilution cryostats meet the demanding requirements and provide continuously sufficient cooling power to cool the solid bodies of samples and necessary measurement equipment. Therefore, the whole sample holder as portrayed in Fig. 3.3 is thermally connected to the base plate of a ^{3}He/^{4}He dilution cryostat.

The following describes how such refrigerators operate in principle and how they must be equipped to suit measurements on Josephson junction devices. In this work, these samples are probed both by DC and RF measurement techniques, whose essentials are presented after that.

3.2.1 Pulse tube precooled ^{3}He/^{4}He dilution cryostat

^{3}He/^{4}He dilution cryostats are multistage cooling units whose precooling to a few kelvins can either be implemented by baths of liquid nitrogen and helium or mechanical cryocoolers like a closed cycle pulse tube refrigerator. The latter are very popular as they circumvent moving parts in their low-temperature part, and thus avoid vibrations. As shown in Fig. 3.4, two of such pulse tube refrigerators are thermally connected to the first two stages and precool the ^{3}He/^{4}He mixture down to liquefaction. This is cooled further by means of a Joule-Thomson valve, where the temperature decreases under isenthalpic expansion [ADK18].

Now, below $\approx$ 870 mK, the ^{3}He/^{4}He mixture undergoes a spontaneous phase transition and separates in two phases: a ^{3}He rich phase (concentrated phase) that consists almost exclusively of ^{3}He and a ^{3}He poor phase (dilute phase) that consists in its thermodynamic equilibrium at 20 mK of $\approx$ 93.5 % ^{4}He and $\approx$ 6.5 % ^{3}He [EH05]. Due to their different densities, the concentrated phase in the mixing chamber floats above the dilute phase, which is connected with a reservoir at still. There, the higher temperature and the decreased pressure favor the evaporation of almost exclusively ^{3}He because of the isotopes' different vapor pressures. As a result, there is a nonequilibrium with a concentration gradient that goes along with an osmotic pressure between the dilute phases at the base and still stage. To compensate for this and maintain the thermodynamic equilibrium, ^{3}He of the concentrated phase must mix itself with the dilute phase. The required entropy of mixing is extracted from the system in the form of heat [EH05]. To ensure continuous cooling, the evaporated ^{3}He is sent to the cryostat again, cooled by heat exchangers on the condenser side, and fed into the concentrated phase in the mixing chamber.

Even though this cooling principle enables theoretically arbitrary low temperatures, its finite cooling power $\sim$ 10 µW limits the base temperature of ^{3}He/^{4}He dilution cryostats to $\sim$ 10 mK, because of thermal connections between the stages, heat input through circulating mixture, thermal radiation, and unavoidable Joule heat input of electronic devices. Nontransparent metallic shields at each stage suppress the latter, where the warmest ones have the greatest bearing according to the Stefan–Boltzmann law $P \propto T^4$. For this reason and for convenience, the used *Bluefors LD250* cryostat abstains from a base shield. Thermal connections between the stages as well as their thermal contact with the environment are reduced by

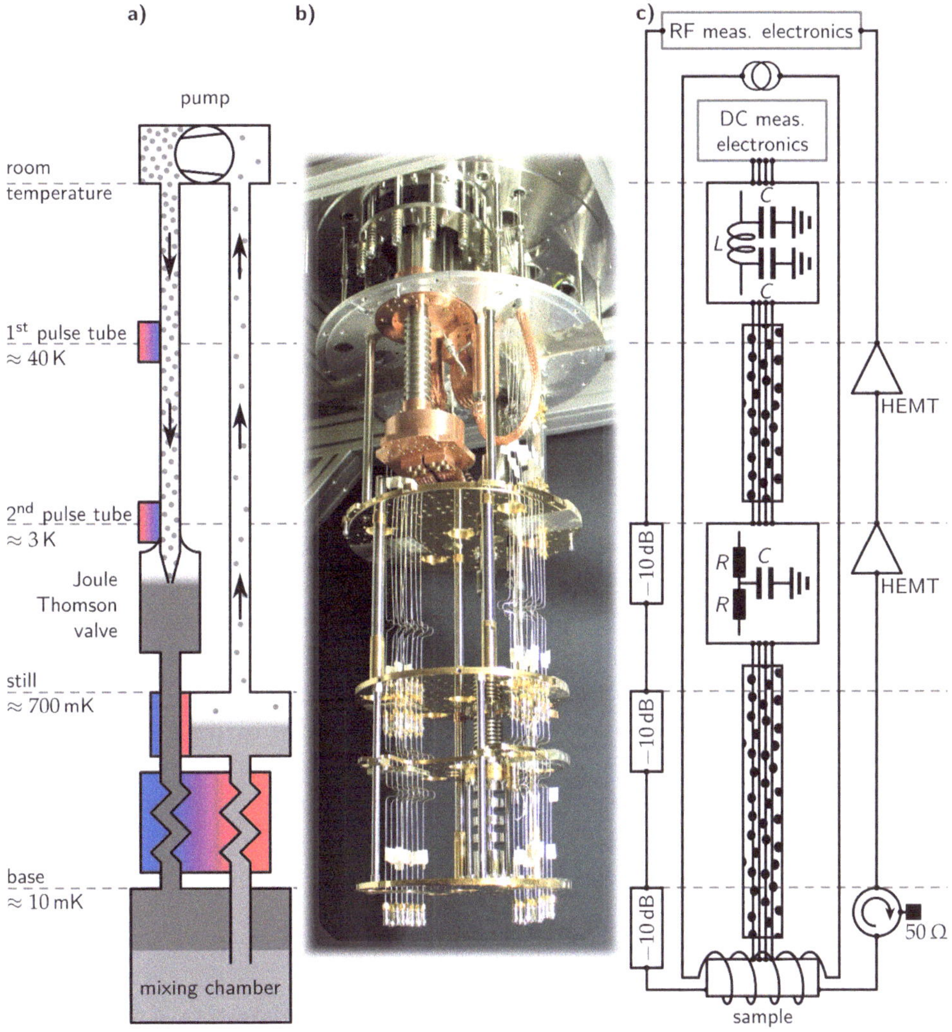

Figure 3.4: Cryogenic measurement setup. a) The cryostat is a multistep refrigerator, whose precooling unit is a two-stage pulse tube generating about 40 K and 3 K, respectively, that is thermally coupled by heat exchangers to the primary dilution refrigerator. By means of a Joule-Thomson valve in this ^{3}He/^{4}He cycle, temperatures at still of about 700 mK are reached, which suffices to obtain a separation into a concentrated and a dilute phase of the ^{3}He/^{4}He and run the dilution refrigeration cycle. Thereby, ^{3}He is pumped down at still from the dilute phase and recondensed to ensure continuous cooling. **b)** The photograph of a *Bluefors LD250* refrigerator shows the implementation used in this work, including the wiring. **c)** The measurement framework consists of low-pass filtered DC wires, coaxial RF lines, and high-current supply lines for magnetic coils.

pumping the cryostat's outer vacuum chamber to $\sim 10^{-6}$ mbar and by an elaborate choice of materials used for the cooling unit and measurement wiring.

The later presented experiments were run in the commercial ^{3}He/^{4}He dilution cryostat *Bluefors LD250*, as depicted by the photograph in Fig. 3.4 b), whose measurement setup is configured to the special needs of Josephson devices. For this purpose, DC and RF connections, as well as high-current supply lines, are mounted, where the heat input to the individual stages due to thermal conductivity is kept minimal by the careful choice of suitable materials and wire cross-sections. Great care is also taken to reduce the thermal noise in the measurement lines by means of filters at different temperatures.

Commercial microwave electronics generate signals at room temperatures with order of magnitudes too high powers for Josephson devices that require a few to a few thousands of photons. This is the reason why thin stainless steel coaxial cables with comparable large damping (23.5 dB m^{-1} at 300 K (compare to Fig. B.1 a) in Appendix B.1) and 14.7 dB m^{-1} at 4 K, both at 1 GHz) can be used as input lines, to reduce the thermal conductance between the individual stages. The stainless steel's comparably high electrical resistivity originates in a strong electron-phonon interaction that is beneficial to thermalize the electrons. For experiments with a few photons, the input signal's noise temperature, which is approximately room temperature, is of particular interest and needs to be reduced. Therefore, it is advantageous to thermalize attenuators at the lowest possible stages that provide sufficient cooling power since each resistor contributes with thermal Johnson-Nyquist noise. The signal's final noise temperature passing an attenuation chain of N stages, each with attenuation factor $a_j > 1$ and temperature T_j (with $T_1 \approx 10\,\mathrm{mK}$ at base and room temperature $T_{N+1} = 300\,\mathrm{K}$) results in [Fri44] [Sch20]

$$T_\mathrm{n} = T_1 + \sum_{i=1}^{N} \frac{T_{i+1}}{\prod_{j=1}^{i} a_j}. \tag{3.3}$$

As the base temperature T_1 contributes without any attenuation, this should be kept as low as possible, and due to the limited cooling power, the attenuation cannot be arbitrarily large there. For this reason, and to decouple the cold stages thermally, attenuators are typically distributed over the three coldest stages, as illustrated in Fig. 3.4 c) [Kri+19]. The total attenuation, which is usually desired to be in the order of ≈ 70 dB, sums up by about 40 dB at 10 GHz in the stainless steel cables and of several attenuators with in total 30 dB in the present setup.

To analyze the signal's few photons, scattered by the sample, at room temperature, resistive coaxial lines are not applicable, by which $\sim 99.99\,\%$ of the signal would be lost. This requires amplification at cold temperatures that is realized by high electron mobility transistors (HEMTs) at the two pulse tube stages.[3] The sample is optimally connected with the first HEMT by superconducting niobium coaxial lines due to their perfect electric conductance and the coexistent thermal isolation. After an amplification of $\approx 35\,\mathrm{dB}$, normal conducting coaxial lines can further be used. In the present setup, the signal is led through silver-plated stainless steel cables that combine weak thermal conductivity in the outer and the middle of the center conductor and good electric conductivity underneath the surface of the center conductors. This silver layer is exactly the area where most of the microwave signals are located according to the skin effect.

In close vicinity to the sample, both input and output signals are low-pass filtered with a cutoff frequency of 18 GHz and infrared filters based on Mie scattering on sapphire powders of different diameters. The latter filter is essential as the coaxial line's and attenuator's typical dielectric PTFE, commonly known by its brand name Teflon, is transparent to infrared radiation. Moreover, to avoid incoming (reflected) signals at the sample's second port, a circulator with a $50\,\Omega$ termination is installed that only allows for transmission in one direction and otherwise attenuates by $\approx 20\,\mathrm{dB}$ (see Fig. 3.4 c)).

Conventional source measure units for transport measurements provide insufficient signal-to-noise ratios, so these signals must also be filtered. In doing so, the DC lines are made from twisted pairs of constantan wires with a rather high resistance of $66\,\Omega\,\mathrm{m}^{-1}$, all embedded in epoxy glue with copper powder acting as infrared filter [LU08], and shielded by a stainless steel mesh. The signals are further low-pass filtered at room temperature by a CLC filter in π topology with a cutoff frequency of about $\approx 300\,\mathrm{kHz}$, and at the second pulse tube stage by a RCR filter in T topology with a cutoff frequency of about $\approx 1\,\mathrm{kHz}$. The latter RCR filter is installed at $\approx 3\,\mathrm{K}$ in order to reduce the resistors' thermal Johnson-Nyquist noise voltage $V_\mathrm{n} = \sqrt{4k_\mathrm{B}TR\Delta f}$, while having sufficient cooling power available. More details such as the transmission spectrum of such DC measurement lines are specified in Fig. B.1 b) in Appendix B.1.

Supply lines for magnetic coils that need to carry high currents consist of massive copper wires between room temperature and the first pulse tube stage, whose thermal load is increased in this manner. Due to its good thermal conductance and the lower cooling power at further stages, such copper wires with large cross-sections are not possible to use there while maintaining millikelvin temperatures

[3] A semiconductor heterostructure provides a two-dimensional electron gas with high electron mobility, such that these amplifiers can operate at a few kelvins, where most other semiconducting amplifiers would not be able to operate.

at base. This is why high-temperature superconductors, featuring zero resistance and good thermal isolation, are used between both pulse tube stages. Below this stage, with about $\approx 3\,\mathrm{K}$, conventional metallic superconductors are available that are way cheaper and much easier to handle. For this purpose, wires are used where multiple filaments of NbTi are embedded in a CuNi matrix. With this combination of materials, avoiding plug-connections and using only clamp connections and superconducting solder joints, the wire resistance is minimized to about $0.3\,\Omega$, while enabling coil currents of a few amperes with negligible Joule heat input to the cryostat.

As ^{3}He/^{4}He dilution refrigerators are conceptualized to operate at millikelvin temperatures, but not around 1 K, an additional stage is thermally weakly coupled to the mixing chamber by thin-walled stainless steel tubes if needed for such temperatures. This extra stage has an individual heater and a separate resistive RuO_2 thermometer, thus allowing for precise temperature control and fast temperature sweeps. This gadget benefits from its small thermal mass and weak thermal connection, so it avoids heating the mixing chamber beyond the critical temperature, where the mixture would vaporize. Thus, temperatures around 5 K can be achieved easily, with base not exceeding 100 mK.

3.2.2 Transport measurement setup

To take IV-characteristics of current biased Josephson junctions with critical currents down to a few nanoamperes, high signal-to-noise ratios are required even at current resolutions in the order of tens of picoamperes. The bias current is generated by means of a home-made voltage biased current source, providing different amplification factors ($2 \times 10^{-2}\,\mathrm{A\,V^{-1}}$ to $2 \times 10^{-6}\,\mathrm{A\,V^{-1}}$). Its core is a reference amplifier consisting of two high-precision monolithic dielectrically isolated FETs *OPA2111*. This current source is supplied by a source-measure unit (SMU), whose output signal is prefiltered by the low-noise preamplifier *Stanford SR560*.

The measured voltage drop across the Josephson junction, in turn, is amplified by a low-noise instrumentation amplifier *INA 105KP* and measured by the SMU. IV-characteristics are recorded by linear staircase sweeps, provided by the SMU, which is in this work either the sourcemeter *Keithley 2636A* or the data acquisition system *NIDAQ USB-6259 BNC*, if the real-time resolution is needed.

A method to take data of transport measurements with customized sequences of sweeps is integrated into the Qkit measurement framework [qki] that is developed in Alexey Ustinov's group at KIT.

3.2.3 Microwave measurement setup

Spectroscopic measurements

Scatter parameters of 2 port networks can be acquired during frequency sweeps for each frequency point employing vector network analyzers (VNA) such as the model *Keysight VNA E5080B*, which was used in this work. In doing so, the complex scattered signal, detected at port 2, is related to the complex signal applied at port 1 to obey the definition of the transmission coefficient S_{21} in Eq. 2.33. The obtained microwave spectrum contains the whole cryogenic RF setup in general but is dominated by the Josephson device in case of proper wiring, as described above. As the receiver's (at port 2) typical intermediate frequency (IF) bandwidth is in the order of 1 Hz–10^4 Hz, such measurements analyze the system in the steady-state, which is sufficient for basic characterizations such as resonances.

As theoretically treated in Sec. 2.3.3, coupled systems are typically characterized via two-tone spectroscopy. Thereby, the second system is driven by an additional tone with power P_{d} and frequency f_{d} that is swept and in the vicinity of the resonant frequency, the second system is populated. This results in a dispersive shift of the first resonator monitored by a VNA. The second tone is generated by a local oscillator (LO), such as the model *Keysight N5173B* in this work, and is added to the VNA's probe signal by means of a directional coupler. The setup is schematically shown in Fig. 3.5 a).

Time resolved measurements

If steady-state measurements are insufficient and time-resolved measurements are needed, a time-domain setup, as schematically shown in Fig. 3.5 b) is used. Commercially available electronics, such as arbitrary waveform generators (AWGs) or digital-to-analog converters (DACs), usually operate with a few gigasamples per second, which is why the required pulses cannot be generated directly with carrier frequencies in the gigahertz range. Instead, only the pulses' envelopes are generated with such devices at a maximum intermediate frequency of 500 MHz and mixed with another carrier frequency generated with an LO, which is at most 20 GHz.

To obtain only a single sideband with this up-conversion, two 90° phase-shifted envelopes are generated and multiplied together with the carrier frequency by IQ mixers to form a complex signal. This way, both manipulation and readout pulses are generated, which are then added with a directional coupler before being sent

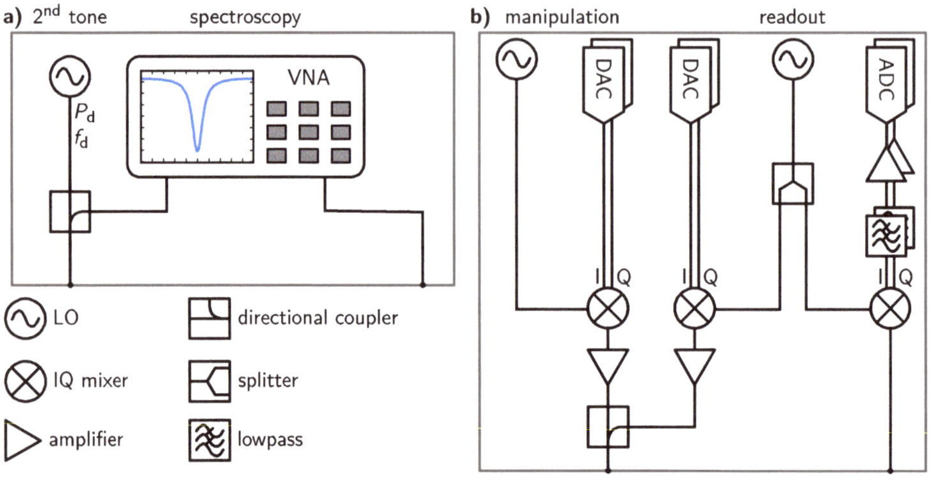

Figure 3.5: Microwave setup for steady-state two-tone spectroscopy and time-resolved measurements. **a)** In a two-tone spectroscopy microwave setup, a probe tone of a vector network analyzer (VNA) with frequency f_{VNA} and power P_{VNA} is added by means of a directional coupler to a drive tone of another local oscillator with frequency f_d and power P_d. The scattered signal is detected at the VNA's receiver and related to the probe signal to determine the scatter parameter S_{21}. **b)** For time-resolved measurements, the pulses' envelopes of both I and Q are generated by a digital-to-analog converter (DAC) and multiplied by IQ mixers with a carrier frequency of a local oscillator (LO) for single-sideband mixing. Both manipulation and readout pulses are amplified with variable gain, and their signals are added by a directional coupler. For the readout signal, the microwave is previously split into equal parts for the up and down conversion to ensure a common reference for both frequency and phase. The signal scattered by the sample is down-converted, low-pass filtered, amplified and recorded using an analog-to-digital converter (ADC). This schematic of the time domain setup is reduced to the basic components. For details, see the thesis of my colleague Richard Gebauer [Geb22], in whose context this setup was developed.

to the sample. The pulses scattered by the sample are again down-converted with IQ mixers using the same LO signal as for the previous up-conversion to ensure a common frequency and phase reference. Before these I and Q measurement signals are recorded with analog-to-digital converters (ADCs), they are low-pass filtered to get rid of high-frequency mixing products and amplified to match the ADC's input levels.

During this work, a time-domain setup was used, which was developed in cooperation with the Institut für Prozessdatenverarbeitung und Elektronik (IPE) at the KIT. The setup combines central processing units (CPUs), field-programmable gate arrays (FPGAs), DACs and ADCs to generate and detect the signals [Geb+20a; Geb+20b]. An in-depth description of the system can be found in [Geb22].

4 Fluxons in high-impedance long Josephson junctions

Long Josephson junctions (LJJs), whose usual bulk electrodes are replaced by thin films of granular aluminum oxide, must be tested experimentally for their suitability for quantum circuits. For this purpose, we experimentally check whether such high-impedance junctions are still governed by the sine-Gordon model, which means that vortices can be excited in this Josephson medium. The existence of vortices in the LJJs is mandatory, which means that the junctions are describable by the sine-Gordon model. Moreover, it needs to be checked whether the additional kinetic inductance can really increase the junction impedance, and thus the quantumness. Furthermore, high-quality tunneling barriers are required to observe quantum coherence, which entails as few dissipative quasiparticle excitations as possible. All these properties are studied in transport measurements by analyzing the LJJs' IV characteristics, their magnetic field dependence, their temperature dependence, and their response to microwave irradiation. The main results of the experiments presented in this chapter have already been published in Ref. [Wil+22].

4.1 Sample considerations and design

In order to study the influence of the kinetic inductance in the junctions' electrodes on the characteristic junction parameters, namely magnetic thickness Λ, Josephson length λ_J, plasma frequency ω_p and Swihart velocity $\bar{c}$, the kinetic inductance of the bottom electrode is varied. For this purpose, the partial oxygen pressure, and thus the sheet resistance, is varied during trilayer deposition and the critical current density is adjusted during static oxidation of the tunnel barrier so that the theoretically predicted Josephson length is a few tens of micrometers. In this manner, junctions can be patterned from the trilayers A, B, and C by photolithography with the same optical masks, varying the width $w = 2\,\mu\text{m}$–$5\,\mu\text{m} < \lambda_J$ and the length $\ell = 20\,\mu\text{m}$–$120\,\mu\text{m} > \lambda_J$ (for stack properties see Tab. 3.1).

Moreover, since the spatial distribution of the phase difference and current density is defined by the junction geometry (see Sec. 2.4.2), three different types of them are studied: In the quasi-overlap geometry, the high-kinetic inductance bottom electrode is contacted along the entire length of the junction (along x) on both sides

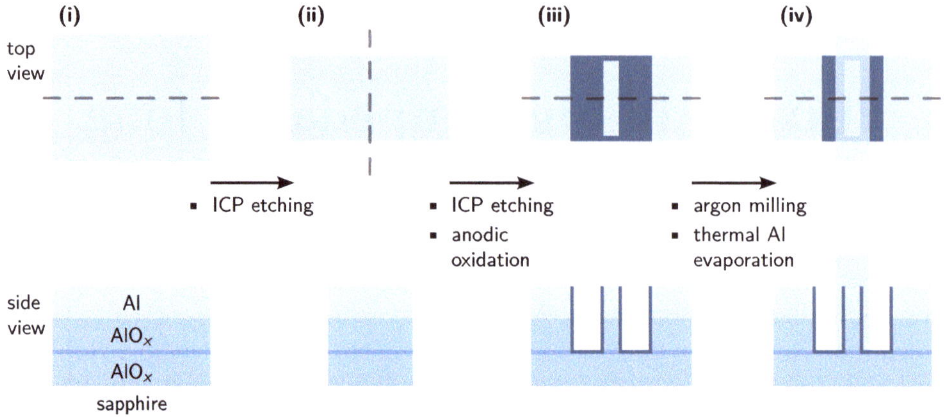

Figure 4.1: Schematic patterning of junction stacks for transport measurements. The upper row illustrates the structure's top view and the lower row its side view along the dashed line after each fabrication step. Starting from the deposited junction stack **(i)** the areas of junctions and DC leads to the bottom electrode are patterned by an argon-chlorine-based inductively coupled plasma (ICP) etching process. The junctions are defined from the remaining stacks **(ii)** by removing the top electrode in the vicinity of the active area by ICP etching. The bare surfaces are further anodic oxidized to insulate the bottom electrode. Onto the resulting active **(iii)** and close parts of the passive region, DC leads are finally contacted to the top electrode by thermal evaporation of pure aluminum, where the galvanic contact is ensured by the previous removal of the native oxide at the surface by argon milling. The quasi-overlap geometry of the final long junction **(iv)**, illustrated here, is characterized by two DC leads to both the bottom and top electrode that allows four-point probes.

with leads that themselves consist only of the AlO_x bottom layer. Their high kinetic inductance compared to the Josephson inductance of the LJJ leads to a homogeneous current distribution since the top electrode is shunted with low-inductive pure aluminum, which is led out at the two short sides of the junctions (along y). This geometry allows four-point probes, even for room temperature precharacterizations. In the inline geometry, bottom and top electrodes are contacted at both ends of the LJJs to allow four-point probes. In this geometry, the bias current typically peaks at the junction ends, and the top electrode's low-inductance shunt increases the current distribution's resulting inhomogeneity since it also becomes asymmetric. In addition, squared short junctions of different sizes are fabricated with a current distribution as homogeneous as possible to determine the critical current density.

4.2 Sample fabrication details

Starting from trilayers A, B, and C that are deposited as described in Sec. 3.1.2, long junctions that provide galvanically connected leads for probing the sample with

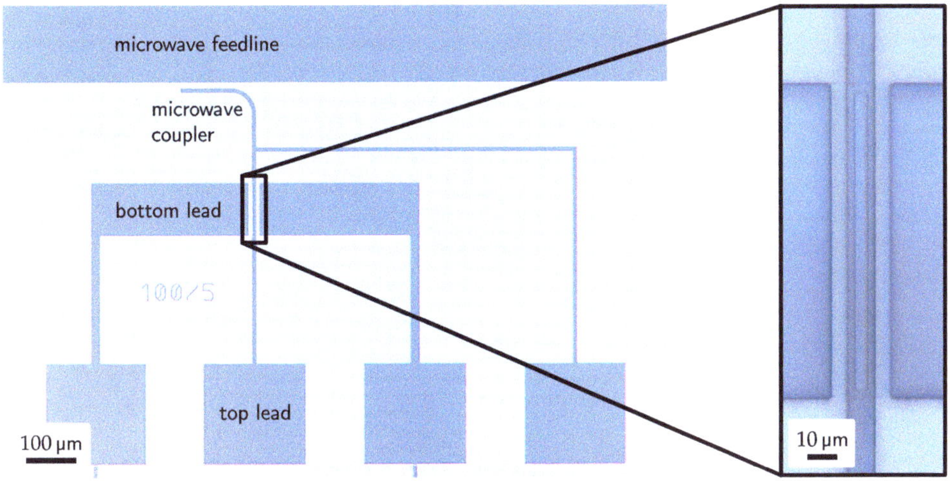

Figure 4.2: Micrograph of a 100 µm × 5 µm long Josephson junction device in quasi-overlap geometry. To enable four-probe measurements at the junction, both its bottom and top electrode are galvanically connected with two bond pads each. One line offers an L-shaped coupler close to a microwave feedline that allows for RF irradiation as required for measuring the plasma frequency. In the inset, the bright field image is superimposed by a dark field micrograph to pronounce edges and highlight the junction underneath the top electrode.

transport measurements, are patterned in three steps: First, the Josephson junctions and their leads to the bottom electrodes are patterned using optical lithography and an argon-chlorine-based inductively coupled plasma (ICP) reactive ion etching (RIE) process. The latter technique offers high etching rates, high selectivity and low-damage processing by combining a high chlorine radical density for the aluminum's chemical etching and a high argon-ion density for physical removal, similar to the sputtering process. In doing so, all parts that are not protected by the photoresist are etched down to the substrate, as illustrated in the top view of Fig. 4.1 (ii).

The junctions' exact dimensions are defined in the second step, where a photoresist covers the active regions and distant parts of the bottom electrode. At the bare areas, only the trilayers' top electrodes are removed by ICP etching, and the resulting surfaces are oxidized using an electrolyte of ammonium pentaborate and ethylene glycol for anodic oxidation[1] (see Fig. 4.1 (iii)). This passivation is mandatory for the subsequent contacting of the top electrodes without shorting the junction stacks by the deposited superconducting layers in the third step.

[1] For the anodic oxidation, all junctions have to be connected to a common pad, where the voltage is applied. These joints between individual junctions are cut during dicing the chip into 5 mm × 5 mm pieces that are installed in the sample holder.

Table 4.1: Properties of the fabricated high-impedance trilayers. The normal conducting sheet resistance of the bottom electrode $R_n^\square$ is extracted from the film deposition. The critical current densities j_c are determined from squared junctions, and Ambegaokar–Baratoff estimations coincide to switching current measurements. The superconducting gap energies are fitted to temperature-dependent gap voltages. Magnetic thickness Λ^{exp} and Josephson penetration depth λ_J are derived from the critical current's magnetic field dependence of inline junctions at $T \approx 300\,\mathrm{mK}$, from which the geometry factor $g(\mathbf{r})$ also follows. From overlap junctions, the Swihart velocity $\bar{c}$ and the impedance Z is acquired from the periodicity of zero-field and Fiske steps at around 1 K. (These data are also published in Ref. [Wil+22].)

trilayer	$R_n^\square$ (Ω)	j_c ($\mathrm{A\,cm^{-2}}$)	Δ_t (μeV)	Δ_b (μeV)
A	38	0.32 ± 0.03	227.1 ± 0.8	278.6 ± 0.6
B	75	12.5 ± 0.3	193.2 ± 0.4	267.5 ± 0.6
C	778	1.90 ± 0.01	201.5 ± 0.7	278.1 ± 0.7

	Λ^{exp} (nm)	λ_J (μm)	$g(\mathbf{r})$ (10^{-2})	$\bar{c}_{zfs}/c_0$ (10^{-3})	$\bar{c}_{FS}/c_0$ (10^{-3})	Z (Ω)
A	69 ± 2	LJJ limit not reached for $\ell \leq 120\,\mu\mathrm{m}$			6.56 ± 0.03	2.79 ± 0.04
B	72 ± 9	17.2 ± 2.2	15 ± 4	4.27 ± 0.06	3.64 ± 0.02	4.11 ± 0.06
C	94 ± 9	19.5 ± 1.8	13 ± 3	3.37 ± 0.08	3.22 ± 0.03	14.0 ± 0.4

To ensure galvanic contact in the active region, the native oxides at the surfaces of the trilayers' former deposited top electrodes are removed by argon milling, whereas the passive region remains insulating. Without breaking the vacuum, the final top electrode and efferent leads are deposited thermally immediately after (see Fig. 4.1 (iv)), which is experimentally much easier to realize with pure aluminum instead of granular aluminum oxide. See Appendix B.2 for details of parameters concerning optical lithography (Tabs. B.1, B.2, and B.4), ICP etching (Tab. B.5), and anodic oxidation (Tab. B.6).

Figure 4.2 shows an optical microscopy image of a typical fabrication results of this process. The sample includes an LJJ in quasi-overlap geometry (see inset) and four DC lines, where one of them is attached to an L-shaped radio frequency (RF) coupler in the vicinity of a microwave feedline.

4.3 IV-characteristics of high-impedance Josephson junctions

First, IV-characteristics of high-impedance Josephson junctions of different sizes and geometries are taken at low temperatures of about 20 mK. As can be seen exemplarily in Fig. 4.3 a), such junctions exhibit a large hysteresis between the switching and retrapping current that confirms a high quality factor originating in only little quasiparticle dissipation. The switching currents themselves are stochastic, which is why their distribution is measured statistically, as depicted in the inset of Fig. 4.3 a). The probability that the junction escapes from the zero-voltage to the nonzero-voltage state within the interval $\mathrm{d}I$ around the current I is given by [FD74a]

$$p^{\mathrm{ecs}}(I)\mathrm{d}I = \Gamma^{\mathrm{ecs}}(I)\left|\frac{\mathrm{d}I_{\mathrm{b}}}{\mathrm{d}t}\right|^{-1}\left(1-\int_0^I p^{\mathrm{ecs}}(\tilde{I})\mathrm{d}\tilde{I}\right)\mathrm{d}I. \tag{4.1a}$$

This probability depends on the inverse bias sweeping rate $\mathrm{d}t/\mathrm{d}I_{\mathrm{b}}$, the activation rate $\Gamma(I)$ and the probability that the junction was not already switched at a lower current. The integral equation for the escape probability, given in Eq. (4.1a), can be solved for the escape probability density (epd) [FD74a]

$$p^{\mathrm{ecs}}(I) = \Gamma^{\mathrm{ecs}}(I)\left|\frac{\mathrm{d}I_{\mathrm{b}}}{\mathrm{d}t}\right|^{-1}\exp\left(-\left|\frac{\mathrm{d}I_{\mathrm{b}}}{\mathrm{d}t}\right|^{-1}\int_0^I \Gamma^{\mathrm{ecs}}(\tilde{I})\mathrm{d}\tilde{I}\right) \tag{4.1b}$$

and further for the escape rate [FD74a]

$$\Gamma^{\mathrm{ecs}}(I) = \left|\frac{\mathrm{d}I_{\mathrm{b}}}{\mathrm{d}t}\right|\ln\left(\frac{\int_I^\infty p^{\mathrm{ecs}}(\tilde{I})\mathrm{d}\tilde{I}}{\int_{I+\Delta I}^\infty p^{\mathrm{ecs}}(\tilde{I})\mathrm{d}\tilde{I}}\right). \tag{4.1c}$$

In the experiment, the bias current is swept with a constant rate $\mathrm{d}I_{\mathrm{b}}/\mathrm{d}t$ using a 16-bit digital to analog converter (DAC) and the switching current from the zero-voltage to the nonzero-voltage state I_{c} is recorded for 10 000 events with a sample rate of 1.25 MS/s by a 16-bit analog to digital converter (ADC) of the data acquisition system *NIDAQ USB-6259 BNC*. A histogram is calculated from these switching currents, by counting the number of switching events n_j^{ecs} within the jth bin of the width ΔI, as done in Fig. 4.4 a). The escape probability density is obtained by normalizing the individual number of events n_j^{ecs} to unity both by the total number of switching events N and the bin width ΔI, so that

$$p_j^{\mathrm{ecs}}(I) = \frac{n_j^{\mathrm{ecs}}}{N\Delta I}. \tag{4.2}$$

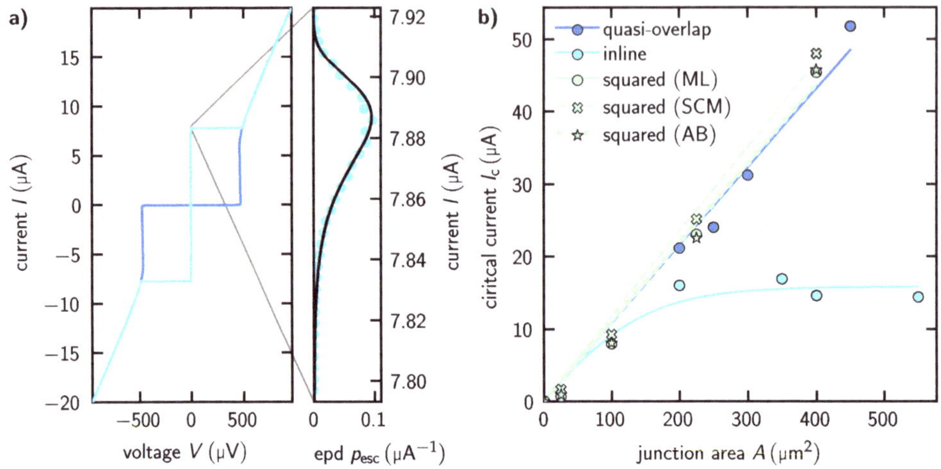

Figure 4.3: Critical currents for different junction sizes and geometries, measured with various techniques. a) The IV-characteristic of a 10 μm × 10 μm junction of trilayer B shows a large hysteresis between critical and retrapping current, indicating low damping, and thus a high junction quality factor. The inset points out the statistics of measured switching currents that differ every time, ramping up the current as each escape from the zero voltage to the nonzero voltage state is a stochastic process. (This figure is also published in Ref. [Wil+22].) **b)** Different methods to determine the critical current are compared exemplarily for square-shaped junctions made from the same trilayer B with various junction areas. The critical current, estimated from the normal state resistance by the Ambeagaokar-Baratoff relation (AB), conforms with the most likely switching current (ML), which is slightly less than the fitted result of switching current measurements (SCM) due to thermal excitation. The area variation of junctions in quasi-overlap and inline geometry is exclusively covered by different lengths ℓ, whereas the width is kept constant at $w = 5$ μm. The critical currents are linear to the junction area for the quasi-overlap and squared geometry, which goes along with homogeneous current distribution. In contrast, the critical current of inline junctions saturates for large junction lengths due to the inhomogeneous current distribution.

As mentioned in Eq. (4.1c), the escape probability determines the junction's activation rate, whose discrete version is calculated from the experimentally acquired values $p_j^{\mathrm{esc}}(I)$ [Wal01] (see Fig. 4.4 a))

$$\Gamma^{\mathrm{ecs}}(I_k) = \frac{\left|\frac{\mathrm{d}I_\mathrm{b}}{\mathrm{d}t}\right|}{\Delta I} \ln\left(\frac{\sum\limits_{j\geq k} p_j^{\mathrm{ecs}}}{\sum\limits_{j\geq k+1} p_j^{\mathrm{ecs}}}\right). \tag{4.3}$$

Within the scope of the RCSJ model (see Sec. 2.4.1), underdamped junctions switch to the nonzero voltage state when overcoming the potential barrier of the junction's tilted washboard potential $U_0(I)$ given by Eq. (2.67). In general, this process can either be because of macroscopic quantum tunneling of the phase through the potential barrier or because of thermal activation over the barrier [DMC85] (see Fig. 4.4 b)).

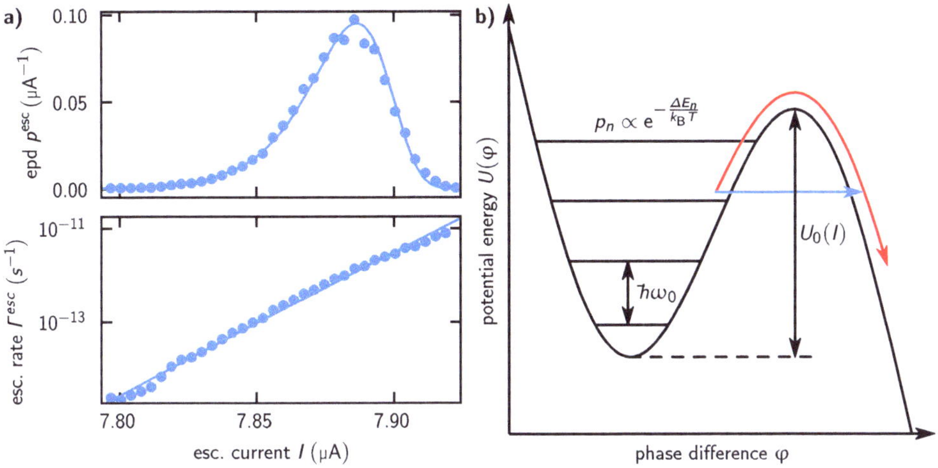

Figure 4.4: Switching current measurements of squared high-impedance junctions. a) To evaluate the distribution of switching currents from the zero-voltage to the nonzero voltage state with 20 000 events, the activation rate is determined from the escape probability distribution (epd), whereby the linear course indicates that this is thermally dominated. A fit of this thermal activation rate as described by Eq. (4.4) provides the critical current I_c and escape temperature T^{esc} as free fit parameters. **b)** The escape of the phase from the junction's potential well can be of thermal and quantum mechanical origin, both depending on the bias-dependent barrier height and the thermal population of higher energy levels.

The crossover temperature, below which the escape rate transitions from thermal to quantum mechanical origin (where it becomes temperature-independent) is given by $T^* = \hbar\omega_p/2\pi k_B$ for underdamped junctions [Aff81; DMC85]. This is not a sharp transition, so even just below the theoretical transition temperature, a relevant number of higher levels are still thermally occupied due to the long tail of the thermal occupancy distribution. Furthermore, it is challenging to bring the electrons in the measurement lines down to the bath temperature so that the real system temperature is higher than the bath temperature [Gia+06].

Accordingly, in the present measurements of the escape from zero voltage to the nonzero voltage state thermal activation must be considered, whose rate is given by the attempt rate $\propto \omega_p$ as well as the ratio of barrier height U_0 and effective escape temperature T^{esc} [BHL83; DMC85]

$$\Gamma^{ecs}_{th}(I) = \frac{\omega_p}{2\pi} \exp\left(-\frac{U_0(I)}{k_B T^{esc}}\right) = \frac{\omega_p}{2\pi} \exp\left(-\frac{E_J \frac{4\sqrt{2}}{3}\left(1 - \frac{I}{I_{c0}}\right)^{3/2}}{k_B T^{esc}}\right). \quad (4.4)$$

The experimentally determined activation rate of Eq. (4.3) is fitted to this thermal activation rate to determine the fitting parameters, namely escape temperature T^* and critical current I_{c0}.

Using square contacts as examples, critical fitted current values I_{c0} (SCM) and most likely switching currents I_c (ML), both determined by switching current measurements, are compared with those estimated by normal resistance according to the Ambegaokar-Baratoff relation of Eq. (2.59) [AB63]. As shown in Fig. 4.3b the Ambegaokar-Baratoff (AB) estimations are in good agreement with the most likely switching currents I_c, which are systematically just below the fitted critical currents I_{c0}.

Figure 4.3 a) also shows the critical currents for different junction geometries, each with different junction areas ℓw, only dependent on ℓ if w is kept constant for long junction geometries. It can be seen that the critical currents of the quasi-overlap junctions increase linearly with the junction's length, coincide with the squared junctions, and thus exhibits a homogeneous current distribution as desired. In contrast, the critical currents of the inline junctions saturate already above a length of 40 µm (corresponding to an area of 200 µm^2 at a fixed width of 5 µm), which allows a Josephson length in the order of 20 µm to be roughly estimated. The critical current densities listed in Tab. 4.1 are determined for squared junctions.

In addition to the critical current, the voltage at which the junction jumps is also an important parameter for IV characteristics of Josephson junctions. After the escape from the zero-voltage to the nonzero voltage state, the phase particle runs down the washboard potential, whereby an equilibrium velocity of energy gain due to the tilt and loss due to dissipation emerges. These losses increase dramatically when the voltage is reached, which is just enough to break Cooper pairs into two dissipative quasiparticles each, so that the equilibrium velocity is approximately equal to this voltage. As shown in Fig. 2.11 c), this voltage corresponds to $V_\Delta = 2\Delta/e$ for SIS junctions with identical superconductors with gap Δ. In case of S_1IS_2 junctions with different superconductors for the bottom and top electrode with gaps Δ_b and Δ_t, respectively, the gap voltage holds $V_\Delta = (\Delta_b + \Delta_t)/e$. In the latter case, a second singularity is possible at $V = |\Delta_b - \Delta_t|/e$ [NSS60; BP82]. To observe these two singularities, the IV characteristics are recorded up to the critical temperature, the two effective gaps are extracted, and if both are measurable the two superconductors' gaps are calculated by

$$\begin{aligned} \Delta_1^{\mathrm{eff}} &= \frac{|\Delta_b - \Delta_t|}{2}, \\ \Delta_2^{\mathrm{eff}} &= \frac{|\Delta_b + \Delta_t|}{2}, \end{aligned} \quad \Leftrightarrow \quad \begin{aligned} \Delta_b &= \Delta_2^{\mathrm{eff}} + \Delta_1^{\mathrm{eff}}, \\ \Delta_t &= \Delta_2^{\mathrm{eff}} - \Delta_1^{\mathrm{eff}}. \end{aligned} \tag{4.5}$$

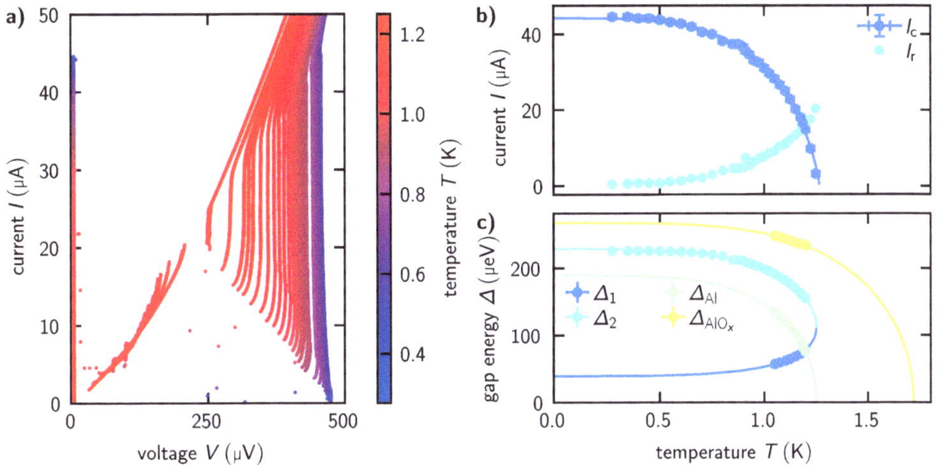

Figure 4.5: Temperature-dependent IV-characteristic and extracted critical current and gap energies. **a)** Since superconductivity is suppressed with increasing temperature, both the gap voltage, which is a measure for the Cooper pair density, and superconductor's overlap, the critical current, decrease and finally disappear completely for $T \to T_c$. The increasingly thermally excited quasiparticles thereby increase the dissipation in the system, which **b)** translates into a higher Stewart-McCumber parameter, thus a larger retrapping current, and **c)** in a reduced superconducting gap.

The resulting gaps of the bottom electrode $\Delta_b \approx 265\,\text{eV}$–$280\,\text{eV}$ and the top electrode $\Delta_t \approx 190\,\text{eV}$–$230\,\text{eV}$ of the three different trilayers A, B, and C corresponds to that of granular aluminum oxide and that of by AlO_x proximitized aluminum, respectively. The temperature dependence of both superconductors can each be described with the Gorter-Casimir two-fluid model [GC34]

$$\Delta_j(T) = \Delta_j(0)\sqrt{1 - \left(\frac{T}{T_{cj}}\right)^4}. \tag{4.6}$$

The gap values for the different trilayers listed in Tab. 4.1 are extrapolated to $T = 0$ from such fits of $\Delta_j(T)$, as for example depicted in Fig. 4.5 c).

4.4 Magnetic diffraction patterns

The high-impedance junction's characteristic lengths along and perpendicular to the tunnel barrier, the Josephson length λ_J and the magnetic thickness Λ, respectively, are experimentally determined from the critical current's dependence on in-plane magnetic fields. As explained in Sec. 2.4.2, externally applied magnetic flux causes a spatial variation of the junction's phase difference along its x-direction, which leads to a diffraction pattern of the critical current intensity. The linear

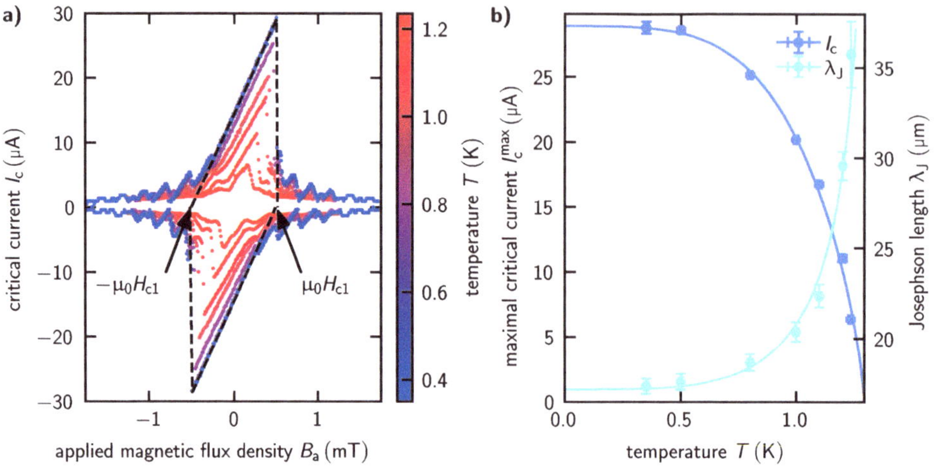

Figure 4.6: Magnetic diffraction pattern of a 110 µm × 5 µm long junction of trilayer B in inline geometry at different temperatures. a) The linear decrease of the Meissner phase confirms reaching the LJJ limit and the extrapolated root of the main lobe corresponds to $\pm H_{c1}$. The asymmetric lobes of the magnetic field-dependent critical current arise because of inhomogeneously distributed bias currents in inline junctions in combination with different inductances in the two electrodes [Sch70; BJV75; Mon+13] (This figure is also published in Ref. [Wil+22].) **b)** The maximum of the main lobe decreases for higher temperatures according to the Ambegaokar-Baratoff model of S_1IS_2 junctions with different superconducting gaps $\Delta_j(T)$, whereas the Josephson length increases because of the temperature-dependent kinetic inductance.

slopes of the lobes in such patterns, as exemplified by Fig. 4.6, confirm that the junction can be considered as long, meaning that self-field effects are not negligible. Accordingly, Meissner-Ochsenfeld-like screening currents can shield the junction's interior from small fields. Since these screening currents penetrate deeper and deeper along the tunnel barrier from the edges, only smaller and smaller additional bias currents can be carried so that the Meissner phase decreases linearly when the magnetic field is increased from zero. A long junction maintains this behavior up to its first critical field [FP63; Sch70]

$$H_{c1} = \frac{\Phi_0}{\pi \mu_0 \Lambda \lambda_J}, \tag{4.7}$$

at which a quantum of flux penetrates the junction by overcoming its edge barrier. Far above, in large in-plane fields, the junction is considered to be completely penetrated by magnetic flux, such that flux quanta are laterally lined up close together along the tunnel barrier. The flux-penetrated extent in the z-direction, the magnetic thickness is accordingly given as

$$\Lambda = \frac{\Phi_0}{\ell \Delta B_a}, \tag{4.8}$$

where ΔB_a is the critical current's periodicity of higher side maxima. As can be seen in Tab. 4.1, L_k affects Λ, since the proximitized top electrode's London penetration depth λ_L enlarges with increasing L_k, whereas the bottom electrode of each stack is without exception in the thin-film limit $d_1 \ll \lambda_{L1}$, and thus contributes to Λ with $d_1/2$ [Wei69].

The first critical field H_{c1} can be determined by the extrapolated zeros of the linear slopes of the main maximum. By inserting this value and the already determined magnetic thickness Λ in Eq. (4.7), the Josephson length can finally be calculated as

$$\lambda_J = \frac{\Phi_0}{\pi \mu_0 H_{c1} \Lambda}. \tag{4.9}$$

Using the definition of the Josephson length from Eq. (2.73), the kinetic inductance, contributing locally to λ_J, can be determined. The comparison of this value $g(\mathbf{r})L_k^\square$ with the kinetic inductance of the bottom layer $L_k^\square$, estimated from the resistance $R_n^\square$ measured *in situ* as $R(d)$ during the sample deposition, yields the geometry factor $g(\mathbf{r})$ on the order of 10^{-1}, as given in Tab. 4.1 [Wil+22].

As can be seen in Fig. 4.6 b), the Josephson length enlarges with increasing temperature since both the critical current and the electrode's dominating kinetic inductance strongly depends on the temperature. The data fit the empirical dependency

$$\lambda_J(T) = \lambda_J(0) \cdot \left(1 - \left(\frac{T}{T_c}\right)^4\right)^{-1/4} \tag{4.10}$$

that is motivated by the temperature dependence of the two-fluid model with the characteristic factor $\left(1 - \left(\frac{T}{T_c}\right)^4\right)^\alpha$. The temperature dependence of the main lobe's maximal critical current is again in good agreement with the model of Ambegaokar and Baratoff of Eq. (2.59) [AB63] as depicted in Fig. 4.6 b).

4.5 Microwave induced switching

The Josephson plasma frequency is the small-amplitude oscillation frequency of a virtual phase particle in a potential well of an unbiased Josephson junction, as illustrated in Fig. 2.10 a). Since the junction's potential is tilted by a bias current, the oscillation frequency strongly depends on the bias as $\omega_0(\gamma) = \omega_p \left(1 - \gamma^2\right)^{1/4}$ [Dah+68] (see Sec. 2.4.1).

An off-resonant external AC bias merely yields weak oscillations of the potential tilt. It thus adds slightly to the DC bias so that smaller DC currents suffice that the junction escapes from the potential well. However, if an additional microwave

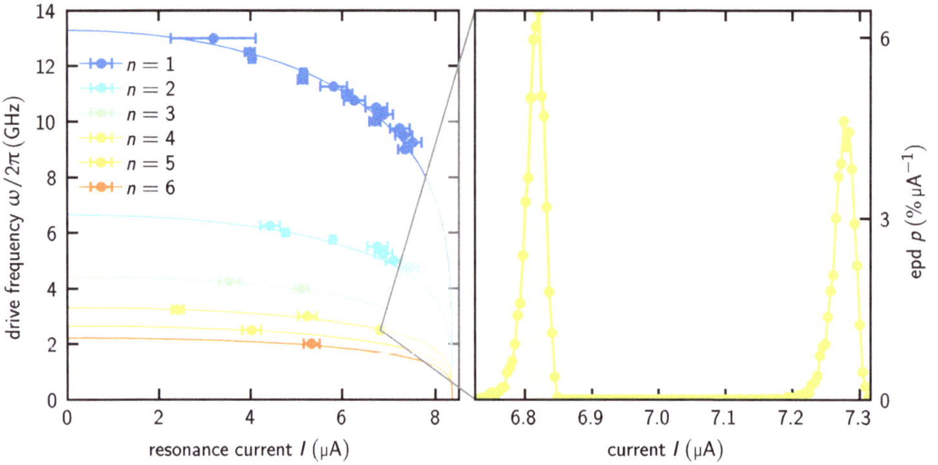

Figure 4.7: Resonance current at different external microwave drives of a squared junction with $(10\,\mu m)^2$ of trilayer B at 20 mK. a) The junction's phase oscillation frequency in a potential well strongly depends on the bias current. Forced oscillations by microwave irradiation can provoke a phase particle's escape from such a potential well if its resonance frequency or n^{th} subharmonics matches the drive frequency. **b)** If so, the junction's switching current becomes multivalued, and the secondary peak in the switching current distribution is identified as resonance current, where its full width half maximum is the error. (This figure is also published in Ref. [Wil+22].)

tone drives the oscillations resonantly, subharmonically or superharmonically, a significantly smaller DC bias current is necessary to escape from a potential well and reach the running state characterized by a nonzero voltage [Grø+04]. This shows up as a secondary peaks in the switching current distribution [Wal+03a; Grø+04; BCG10], as exemplified in Fig. 4.7 b), that is identified as resonant current, for which the fixed external drive frequency equals the junction's resonance frequency ω_0, its integer multiples or fractions of ω_0.

In the experiment with a $10\,\mu m \times 10\,\mu m$ squared junction of trilayer B as an example [Wil+22], switching current measurements with 10 000 events are taken under microwave irradiation of a variable but fixed frequency. After finding a multivalued switching current distribution, the power of the microwave excitation is adjusted so that the additional peak is still clearly visible but does not yet shift towards smaller currents [Wal+03a]. This is the case if the integrals over the individual peaks are nearly equal.

The current value of the additional peak and the corresponding frequency are a single data point in Fig. 2.10 b) and for many such measurements branches of the different harmonic n can be found, obeying

$$\omega_n(\gamma) = \frac{\omega_\mathrm{p}}{n}\left(1 - \left(\frac{I}{I_\mathrm{c}}\right)^2\right)^{1/4}. \tag{4.11}$$

The extrapolation of the oscillation frequency with orthogonal distance regression to $I = 0$ yields the plasma frequency $\omega_\mathrm{p}/2\pi = (13.28 \pm 0.05)\,\mathrm{GHz}$ and the critical current $I_\mathrm{c} = (8.36 \pm 0.08)\,\mathrm{\mu A}$.

This fit enables the calculation of the tunnel barrier's specific capacitance using the definition of the Josephson plasma frequency in Eq. (2.64). The value $c_\mathrm{J} = (36.4 \pm 0.4)\,\mathrm{fF\,\mu m^{-2}}$ is a typical number for tunnel barriers made from thermally oxidized aluminum [Mae+95] and indicates that the additional kinetic inductance does not affect the plasma frequency to the first order.

4.6 Current steps

Apart from the static measurements on long junctions in applied magnetic fields to determine the Josephson length and the measures on short junctions to determine the plasma frequency, their product, the Swihart velocity $\bar{c} = \lambda_\mathrm{J}\omega_\mathrm{p}$, is determined in independent measurements separately. As explained in Sec. 2.4.3 extended Josephson junction stacks can be viewed as nonlinear transmission lines through which different excitations, such as plasma waves or solitons in the form of Josephson vortices, can propagate with at most the Swihart velocity $\bar{c}$ [Ped86]. The impedance mismatch at the two open ends of a one-dimensional long junction forms a cavity that allows standing wave excitations.

One-dimensional linear long junctions with open boundaries have no topological constraints on the number of inherent vortices and antivortices. Therefore, fluxons can be trapped in such junctions even in zero magnetic fields,[2] where the overlap junction's boundaries hold $(\varphi_{\bar{x}} + \beta\varphi_{\bar{x}\bar{t}})\Big|_{\bar{x}=0} = (\varphi_{\bar{x}} + \beta\varphi_{\bar{x}\bar{t}})\Big|_{\bar{x}=\ell} = 0$. In this geometry, solitons are driven shuttle-like along the tunnel barrier by a Magnus-Lorentz force caused by a nearly homogeneously distributed bias current. At the junction ends, they are reflected, reversing not only their direction but also their polarity to

[2] At first sight this may seem counterintuitive, but theoretical considerations show that in long junctions even the smallest perturbations are sufficient for a Josephson vortices to be energetically more favorable [Lik86].

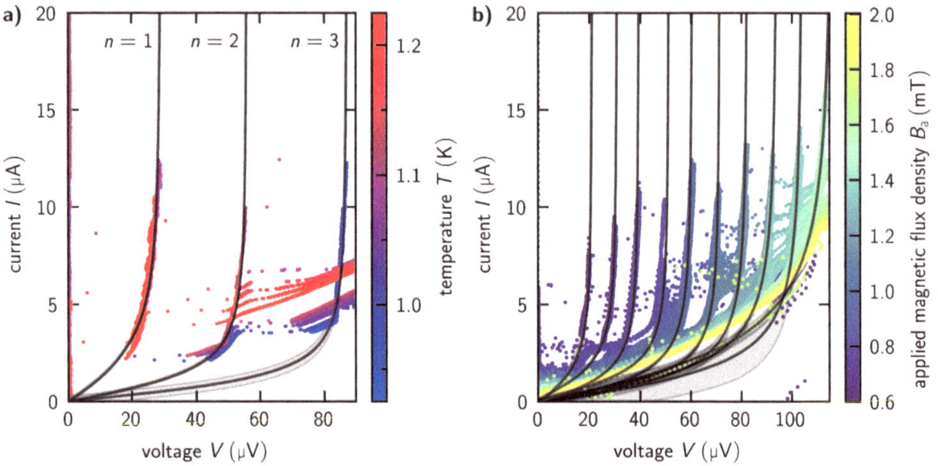

Figure 4.8: Current singularities due to junction cavity resonances of 100 μm × 5 μm and 110 μm × 5 μm long junctions of sample B in quasi-overlap geometry without and with magnetic fields, respectively. a) Zero-field steps arise only with sufficient damping in the junctions, which is realized by temperatures just below the critical temperature $T_c \approx 1.25\,\mathrm{K}$. **b)** Fiske steps occur at different magnetic fields and their characteristic rounded shape originates from the increased damping at $T = 1.0\,\mathrm{K}$. Both current steps feature a bias current depending characteristic shape according to Eq. (4.12), whose fits and errors as displayed as dark gray line and the gray shaded area. (These figures are also published in Ref. [Wil+22].)

satisfy the boundary condition. This process can be regarded as a virtual collision with an antisoliton. Thereby, soliton and antisoliton move towards the junction boundary with opposite velocities, collide at the edge and pass through each other if their velocities are large enough, such that the soliton leaves, whereas the antisoliton enters the junction [Ust05]. Due to its reverse polarity, this antisoliton is driven in the opposite direction of the soliton. In doing so, the soliton is reflected as antisoliton at the one junction boundary and vice versa at the opposite boundary, yielding resonances in the junction cavity in zero magnetic fields [CFL71; FD73; FD74b].

Accordingly, two 2π-kinks pass each point of the junction per period Δt, giving rise to the time-averaged voltage $\langle V \rangle = 2\Phi_0/\Delta t$ across the junction. The period is given by the junction length and the propagation velocity from Eq. (2.98) as $\Delta t = 2\ell/v$, so that the voltage drop is given for $\beta = 0$ by [MS78; BP82; Lik86]

$$V_n^{\mathrm{ZFS}}(\gamma) = \frac{\Phi_0 \bar{c}}{\ell} \frac{n}{\sqrt{1 + \left(\frac{4\alpha}{\pi\gamma}\right)^2}}. \tag{4.12a}$$

Here, n denotes the number of participating solitons that cause equidistant current steps in the IV curves, which approach the voltage $\Phi_0\bar{c}/\ell$ in the relativistic limit

($v \to \bar{c}$) achieved for large bias current. For their experimental observation, homogeneous bias current distributions such as those obtained in the quasi-overlap geometry are advantageous to ensure a shuttle-like soliton motion and avoid their edge pinning as observed in inline junctions.

To reliably access such current singularity features in the IV characteristic arising from the subgap resistance branch, it further helps to increase the dissipation of the underdamped junctions by increasing the sample temperature. Then, however, as shown in Fig. 4.8 a), the Stewart-McCumber branch cuts the lower part of the higher-order steps. Nevertheless, the current steps are fitted to Eq. (4.12) to determine the Swihart velocity.

In finite magnetic fields, this perception of shuttle-like vortex motion remains valid only in a very narrow parameter range [EFP83] of low magnetic fields $H \approx H_{c1}/2$, intermediate junction lengths $\bar{\ell} \lesssim 10$, and comparable large damping [EFP83]. In this configuration, fluxons can travel the Josephson transmission line without being reflected at the junction edges, but being rather annihilated at a boundary. The released energy travels back as a slowly decaying plasmon, which in turn triggers a fluxon at the opposite boundary [Due+81; Ust05]. Consequently, a single 2π per fluxon kink passes each point of the junction per full period, which results in half the voltage of the zero-field steps [Fis64; Kul65; Kul67]

$$V_n^{\mathrm{FS}}(\gamma) = \frac{\Phi_0 \bar{c}}{2\ell} \frac{n}{\sqrt{1+\left(\frac{4\alpha}{\pi\gamma}\right)^2}}. \tag{4.12b}$$

Thus, such Fiske steps with even numbers of participating fluxons n coincide exactly with the former zero-field steps. Fiske steps can also be observed apart from this narrow parameter range with identical characteristics, although the underlying mechanism differs. In general, the Josephson frequency of a biased junction excites electromagnetic standing waves in the junction cavity, whose higher harmonics give rise to periodic current singularities, as shown in Fig. 4.8 b). For underdamped junctions, cavity oscillations are unstable for $\omega \lesssim \omega_{\mathrm{p}}$ [Cir+97; Cir+98], which explains missing the first Fiske step in Fig. 4.8 b).

Nevertheless, the Swihart velocity $\bar{c}$ is determined by the periodicity of the current singularities and the well-defined junction length ℓ (see Tab. 4.1). Knowing the junction width w from fabrication and the specific capacitance c_{J}, the impedance of the Josephson medium can be calculated by $Z_{\mathrm{J}} = \frac{1}{w\bar{c}c_{\mathrm{J}}}$

To analyze the impact of the electrode's kinetic inductance on LJJs, their characteristic parameters, listed in Tab. 4.1, are compared with estimations for conventional LJJs with equal tunnel barrier properties j_{c} and c_{J}, but made from pure aluminum.

Here, we assume that pure aluminum electrodes have negligible kinetic inductance. The result of this comparison is that the electrode's kinetic inductance reduces both the Josephson length λ_J and the Swihart velocity $\bar{c}$ by a factor of up to 40, while the Josephson plasma frequency ω_p remains nearly unchanged. Accordingly, the wave impedance of LJJs is increased by the same factor. The inductance contributing to Josephson plasma oscillations is dominated by the macroscopic stack tunnel barrier rather than the nanoscopic tunnel barriers in AlO_x due to the much stronger intergrain coupling so that the increase in L_k can be neglected to the first order. The combination of the independently measured parameters corresponds to the conventional sine-Gordon model with $\bar{c} = \lambda_J \omega_p$.

4.7 Concluding remarks

The reported transport measurement results demonstrate that the novel high-impedance Josephson junctions made from disordered oxidized aluminum as a high-kinetic inductance superconductor are promising candidates for studying the quantum coherence of Josephson vortices. These long junctions feature high quality. The observation of zero-field steps proves the existence of Josephson vortices in the long junctions, and thus confirms the sine-Gordon model being applicable for high-impedance long junctions.

Beyond, the experiments demonstrate a decrease in the vortex's size and a reduction of its limiting (Swihart) velocity by about one order of magnitude compared with conventional LJJs, whereas the plasma frequency is unaffected to the first order. The measured Swihart velocities down to a small fraction of 3×10^{-3} of the vacuum light velocity, in turn, correspond to an increase in the junction's wave impedance up to $14\,\Omega$ compared to $4\,\Omega$ of conventional, similarly made LJJs. The enhancement of this impedance goes along with an increased quantumness $\bar{\hbar}_J \propto Z$. As $\bar{\hbar}_J$ plays the key role in experimentally reaching the quantum regime of Josephson vortex dynamics [Wal+03b], high-kinetic inductance electrodes also facilitate observing the quantum electrodynamics phenomena in long Josephson junctions.

Apart from the goal in this work of quantum coherence of Josephson vortices, these high-impedance junctions are promising for the research field of active Josephson microwave components. The high-kinetic inductance electrodes enable tailoring the junction impedance and facilitate solving the long-standing problem of impedance matching LJJs to external circuits and $50\,\Omega$ cables. Matching the impedance to external loads is crucial for increasing the efficiency of Josephson flux-flow oscillators used for microwave generation and amplification [Wil+22].

5 Long junctions embedded in a high-impedance resonator

Observations of quantum coherence are experimentally challenging using DC measurements since the sample is galvanically connected to the environment, representing a massive decoherence channel. However, excitations with alternating electromagnetic fields allow for manipulating and readout the circuit without such DC lines. For this purpose, long junctions are coupled to high-quality microwave resonators that are completely galvanically decoupled from the environment and act as very narrow band-pass filters even for AC by coupling only on resonance. This readout scheme is well established for superconducting quantum bits [Bla+04; Wal+04], but not yet for long Josephson junctions. The following chapter elaborates on a method to couple high-impedance long Josephson junctions (LJJs) to such resonators and their fabrication process. Further, spectroscopy measurement data involving the anharmonicity and the magnetic field dependence are compared to a theoretical model of coupled junction and resonator.

5.1 Sample considerations and design

One possible way to couple a high-impedance long junction with a distributed $\lambda/2$ high-kinetic inductance resonator is to expand the junction's bottom electrode to a resonator. In other words, a section of the resonator is shared as the bottom electrode of the long junction, where a tunnel barrier and a top electrode are deposited on top (see Fig. 5.1).

For infinitely long junctions, the currents in this assembly, driven by alternating electromagnetic fields, branch out in the two electrodes according to their inductance ratio as described by Kirchhoff's current law. Because of the supercurrent tunneling through the junction, a resonator-junction interaction is achieved, whose coupling strength is controlled by the inductance ratio of the bottom to the top electrode as well as the widths of the junction and resonator. Strong coupling can, for instance, be realized by a high-kinetic inductance bottom electrode made of AlO_x and a top electrode of pure aluminum, whose orders of magnitude smaller inductance is negligible (see Fig. 5.1). In this extreme case, the major part of the

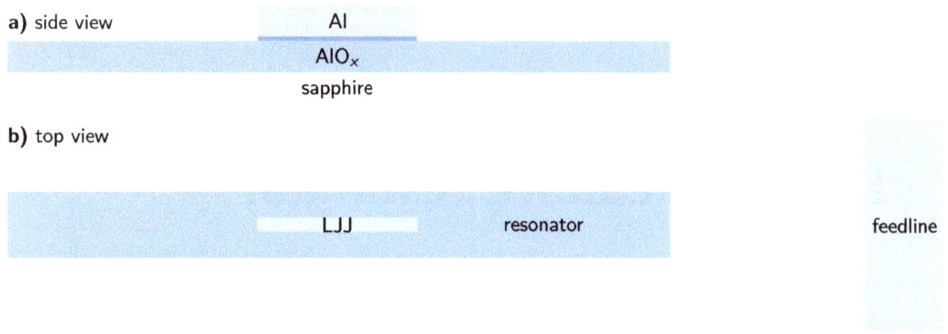

Figure 5.1: Schematic sample design for long junctions embedded in a microwave resonator. a) A open-end microstrip *LC* transmission line forms a distributed $\lambda/2$ resonator. The center part of this high-kinetic inductance resonator serves as bottom electrode of a high-impedance long Josephson junction, whose top electrode is made from a pure aluminum. **b)** The resonator is coupled capacitively to a microwave feedline in notch type configuration.

current driven in the resonator tunnels through the junction and flows in the upper branch.

The extent of the junction's edge regions, where this current tunnels across the barrier, is governed by the ratio between Josephson and electrode inductance, and thus by the dimensionless screening parameter $\beta_L = L_0/L_J$, which is related to the Josephson length.

As distributed resonators exhibit spatially varying electromagnetic field distributions, the choice of the LJJ's position is decisive: Assuming a homogeneous kinetic inductivity, the spatial center corresponds to the symmetry point of the current distribution in the resonator. This means there are current nodes for even and antinodes for odd harmonics, respectively.

For this reason, the LJJ's impact on the resonator varies depending on the excited mode since the phase drop along the junction $\Delta\phi \propto LI$ differs.

FEM simulations[1] analyze the impact of varying inductance in the resonator's center part (this is where the LJJ will be placed) on its resonant frequency. In the extreme case of zero inductance of the inner 10 % of the resonator area even harmonics are affected insignificantly as the resonant frequency changes only by 0.03 % for the second harmonic and 0.8 % for the fourth harmonic. In contrast, odd harmonics with current antinodes in the center of the microstrip resonator

[1] The resonant frequencies are estimated by employing *Ansys HFSS*, a multipurpose, full-wave 3D electromagnetic simulation software that utilizes a finite element method (FEM) eigenmode solver.

drastically change their resonant frequency by $\approx 10.4\,\%$. Thus, spectroscopy measurements to determine higher-order resonance frequencies of such samples can be used to experimentally verify whether the proposed arrangement couples LJJs to resonators, making the former suitable for RF readout.

As LJJs' inductances are sensitive to magnetic in-plane fields, such fields should significantly impact the resonant frequencies. For this purpose, a model of phase-biased junctions was developed in this work, which is described in detail in Sec. 5.3.

To attribute the expected anharmonicity and modified magnetic field dependence to the LJJ, a reference resonator without any LJJ on top is fabricated in the same fabrication run. Frequency multiplexing of several notch-type resonators on a common feedline is utilized to measure multiple resonators with a small spatial distance to be as insensitive as possible to inhomogeneities in the trilayer. In addition, fewer RF lines and cooldowns are needed to measure and compare different resonators, but the measured resonances need to be unambiguously identified.

Two-dimensional on-chip waveguides are usually realized in a coplanar structure, where center conductors are separated by an insulating gap from semi-infinite ground planes or by microstrip lines, whose ground electrodes are placed on the chip's backside. Since homogeneous in-plane fields are applied in the present experiment, the microstrip variant with a normal instead of superconducting ground plane will be used. Normal metals prevent bending magnetic fields near the junctions due to Meissner screening in the ground plane. Besides, a normal conducting backside metallization avoids trapping Abrikosov vortices, which would interfere with the applied magnetic field.

In the AC experiments, it is convenient to use microwave frequencies on the order of a few gigahertz since their corresponding energies are, on the one hand, sufficiently below the superconducting gap (e.g., $2\Delta/h \approx 87\,\mathrm{GHz}$ for aluminum) to avoid quasiparticle excitations. On the other hand, gigahertz frequencies are adequately above the available temperatures ($k_\mathrm{B}T/h \approx 200\,\mathrm{MHz}$ for conventional dilution refrigerators reaching $T \sim 10\,\mathrm{mK}$) to dominate thermal excitations and enable deliberate manipulation. Moreover, the manifold commercial electronic equipment used for telecommunication operates in the kilohertz to gigahertz band. Access to this microwave technology required for the RF measurements, such as vector network analyzers, microwave sources, amplifiers suitable for cryogenic temperatures, circulators, mixers, etc., ease experimental implementations.

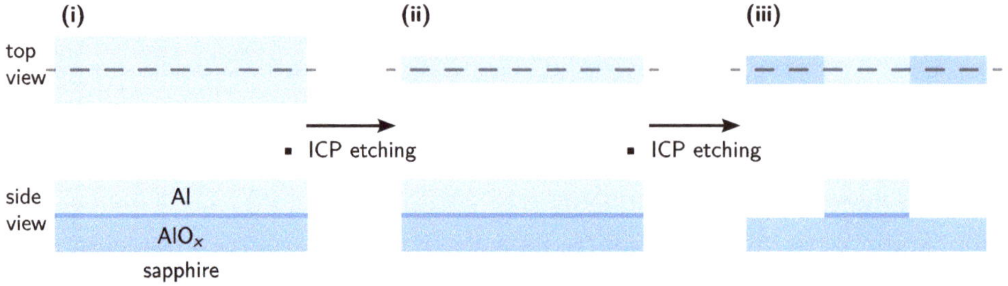

Figure 5.2: Schematic patterning of a long junction embedded in a microwave resonator. The upper row depicts the circuit's top view, and the lower row the side view along the resonator and long junction (dashed line) after each fabrication step. Starting from the deposited junction stack **(i)** the resonator area that includes the junction is patterned by an argon-chlorine-based inductively coupled plasma (ICP) etching process. The top electrode of the remaining trilayer **(ii)** is also removed by ICP etching at the outer parts of the resonator to define the long junction in the middle of the resonator only **(iii)**.

To achieve a large coupling strength between resonator and junction, granular aluminum oxide with $R_\mathrm{n}^\square = 2\,\mathrm{k\Omega}$ which corresponds to $L_\mathrm{k}^\square = 1.5\,\mathrm{nH}$, as bottom and pure aluminum as top electrode are desired. In order to exhibit fundamental modes in the frequency range 4.5 GHz–7.5 GHz, the resonators' lengths are designed to $l \approx 300\,\mu\mathrm{m}$–$500\,\mu\mathrm{m}$ at a uniform width of $w = 2\,\mu\mathrm{m}$. The junctions extend over the entire width of the resonators $w = \mathrm{w}$ and their lengths are fixed to 10 % of the resonator length, so that $\ell \approx 30\,\mu\mathrm{m}$–$50\,\mu\mathrm{m}$. While these values should correspond to long normalized junction lengths $\bar{\ell} \approx 9$–15, the Josephson length is targeted around $\lambda_\mathrm{J} \sim 3.5\,\mu\mathrm{m}$. Therefore, the desired critical current density is estimated to be $j_\mathrm{c} \sim 100\,\mathrm{A\,cm^{-2}}$.

5.2 Sample fabrication details

The trilayer D, whose measurements are shown below, meets the desired parameters (see Tab. 3.1) and is sputtered as described in the Sec. 3.1.2. Resonators and long junctions are patterned each by photolithography and argon-chlorine-based inductively coupled plasma (ICP) reactive ion etching (RIE), as schematically illustrated in Fig. 5.2: First, areas of the resonators, including the long junctions, are covered by a photoresist, and the rest of the trilayer is subsequently etched down to the sapphire substrate. Second, only the smaller junction area in the center of the resonator is covered by a photoresist, and the top electrode layer only is removed next to it so that the outer part of the resonator consists only of the AlO$_x$ layer. This step is delicate because it must be ensured that the junction, i.e., the low-inductance aluminum top electrode, is completely removed while the plasma does not etch too deeply into the bottom electrode. Unfortunately, since there is no selective etch that

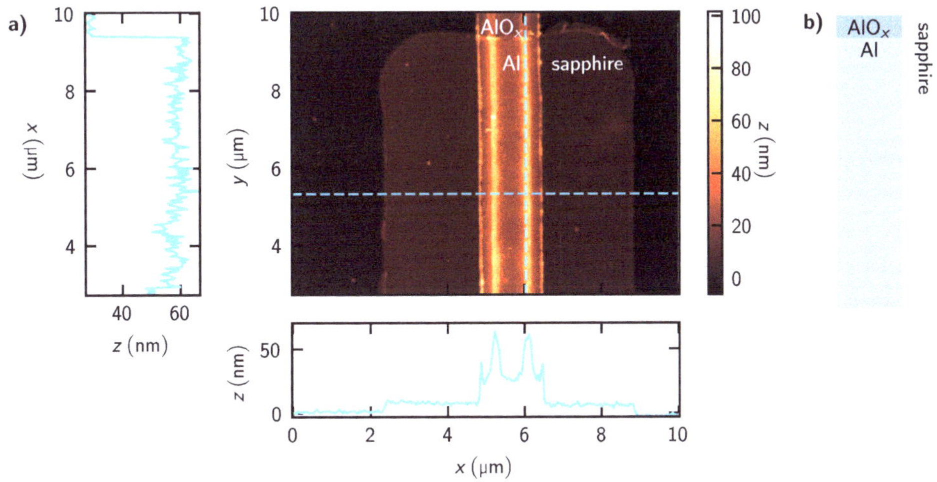

Figure 5.3: Atomic force micrograph of a AlO_x strip with an LJJ on top. The atomic force micrograph (**a)**) shows the long junction embedded in a microstrip resonator, and the schematic of the microstructure is illustrated in **b)**. From the measured height profile, the width of the vertical trilayer strip is estimated to $\approx 1.7\,\mu m$. In the lower part ($y \lesssim 9.5\,\mu m$), the long junction's trilayer stack survives, whereas the top electrode is etched away in the upper region. For this purpose, a photoresist, which protects against etching, is patterned photolithographically with a horizontal excess length of $\approx 2\,\mu m$, resulting in a slightly higher substrate there.

etches only pure aluminum but not granular aluminum oxide, this etching result can only be achieved by choosing the appropriate etching time. See Appendix B.2 for details of parameters concerning optical lithography (Tabs. B.1, B.2, and B.4) and ICP etching (Tab. B.5).

Finally, an additional 100 nm thick layer of niobium is deposited onto the microstrip feedline in an optical liftoff process. Niobium features high critical temperature, high critical magnetic in-plane fields, and negligible kinetic inductance that allows for well-defined impedance matching to the standard microwave components with $50\,\Omega$. To confine the electromagnetic fields properly, a backside metallization is deposited in the form of a 100 nm thick layer of silver with a 5 nm thick coat of titanium beneath to improve its adhesion on the sapphire substrate.

During patterning of sample D, the top electrode was slightly overetched, that is to say the granular aluminum layer was thinned down by a few nanometers. As the thickness of the resonator layer decreases, the sheet resistance or rather the kinetic inductance increases (compare the $R_n(d)$ curve in Fig. 3.2 b)), which can be verified by test structures at room temperature. Four-point probes on test strips with 150–250 squares yield a sheet resistance of $R_n^\square = (4.6 \pm 0.6)\,k\Omega$, which corresponds to $L_k^\square = (4.9 \pm 0.6)\,nH$ according to Eq. (2.32). FEM simulations with this estimated kinetic inductance value yield fundamental modes between 2.6 GHz–4.3 GHz for the six fabricated resonators, whose dimensions are listed in Tab. C.1 in Appendix C.

The height profile from atomic force microscopy of a typical fabrication results of this process is depicted in Fig. 5.3. There, the structures can be measured in all three spatial directions so that the width of the resonators and junctions is determined to be about $\approx 1.7\,\mu\text{m}$. From the height profile, it can be confirmed that a thin layer of the bottom electrode was removed in the second etch step. The window extending beyond the junction comes from the fact that the resist in the second etch step protrudes about $2\,\mu\text{m}$ beyond the functional structure onto the bare substrate to become insensitive to misalignment along this direction.

5.3 Phase-biased Josephson junctions

To determine the eigenfrequencies of the coupled system of the long junction and microwave resonator, the eigenmode solver of the electromagnetic FEM simulation suite *Ansys HFSS* is deployed. For this purpose, the inductances of the resonator's parts with and without the junction are treated differently. In the outer regions of the bare AlO_x layer, the dominant kinetic inductance is estimated from Eq. (2.28). The spatial phase distribution along the junction is analyzed in the resonator's critical center part with the LJJ. The inductance, included in the FEM simulation, is then calculated from the phase drop between the two junction ends.

For this analysis of the phase distribution, the inductance in the top electrode L_0^{Al} is neglected since it is small compared to the inductance in the bottom electrode $L_0^{\mathrm{AlO}_x} \gg L_0^{\mathrm{Al}}$. As shown in Fig. 5.4, the distributed LJJ is divided into $N \to \infty$ discrete segments and Kirchhoff's current law is applied at each node $j \in [1, N]$. In doing so, inside the junction, the jth node flux is coupled by the nearest neighbors as

$$\frac{\Phi_{j+1} - \Phi_j - \Phi_{j+1}^{\mathrm{ext}} + \Phi_j^{\mathrm{ext}}}{L_0/(N-1)} + \frac{I_\mathrm{c}}{N}\sin\left(\frac{2\pi}{\Phi_0}\Phi_j\right) = \frac{\Phi_j - \Phi_{j-1} - \Phi_j^{\mathrm{ext}} + \Phi_{j-1}^{\mathrm{ext}}}{L_0/(N-1)}. \tag{5.1a}$$

At the junction edges, the bias current I, driven by the resonator, are bought into the equation, so that the boundary conditions hold

$$\frac{\Phi_2 - \Phi_1 - \Phi_2^{\mathrm{ext}} + \Phi_1^{\mathrm{ext}}}{L_0/(N-1)} - \frac{I_\mathrm{c}}{N}\sin\left(\frac{2\pi}{\Phi_0}\Phi_1\right) = I, \tag{5.1b}$$

$$I - \frac{I_\mathrm{c}}{N}\sin\left(\frac{2\pi}{\Phi_0}\Phi_{N-1}\right) = \frac{\Phi_N - \Phi_{N-1} - \Phi_N^{\mathrm{ext}} + \Phi_{N-1}^{\mathrm{ext}}}{L_0/(N-1)}. \tag{5.1c}$$

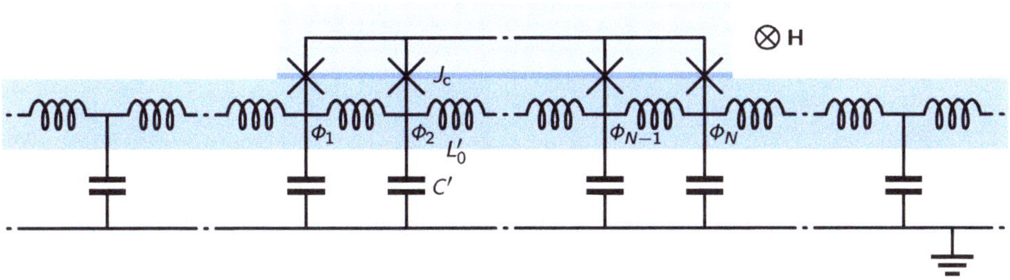

Figure 5.4: Schematic circuit of a galvanically coupled long junction to a microwave resonator. The microstrip structure forms an open-end LC transmission line with the strip's geometric and kinetic inductance $L_0 = L'_0 l$ and the capacitance against the backside ground electrode $C = C'l$. The resonator's conductor is shared as the long junction's bottom electrode, which is weakly coupled with the top electrode by an insulating tunnel barrier with the critical current $I_c = J_c \ell$. Since the bottom electrode's kinetically dominated inductance exceeds the purely geometric inductance of the top electrode by orders of magnitude, this is dropped in the schematic circuit.

The node fluxes can be translated into node phases by means of the junction's well-known flux-phase relation $\Phi_j = \frac{\Phi_0}{2\pi}\varphi_j$. Additionally, a homogeneous external magnetic flux causes a constant external phase drop between two adjacent nodes $\Phi^{\text{ext}}_{j+1} - \Phi^{\text{ext}}_j = \frac{\Phi_0}{2\pi}\frac{\varphi^{\text{ext}}}{N-1} = \frac{\Phi_0}{2\pi}\Delta\varphi^{\text{ext}}\ \forall\ j$. By further identifying the Josephson inductance $L_\text{J} = \frac{\Phi_0}{2\pi I_c}$ and introducing the screening parameter $\beta_\text{L} = \frac{L_0}{L_\text{J}} = \frac{2\pi I_c L_0}{\Phi_0}$ (for details see Appendix A.2), the conditional equations can be formulated as

$$\varphi_2 = \varphi_1 + \Delta\varphi^{\text{ext}} + \frac{\beta_\text{L}}{N(N-1)}\sin(\varphi_0) + \frac{\beta_\text{L}\gamma}{N-1}, \tag{5.2a}$$

$$\varphi_{j+1} = 2\varphi_j - \varphi_{j-1} - \frac{\beta_\text{L}}{N(N-1)}\sin(\varphi_j), \tag{5.2b}$$

$$\frac{\beta_\text{L}\gamma}{N-1} = \varphi_N - \varphi_{N-1} - \Delta\varphi^{\text{ext}} + \frac{\beta_\text{L}}{N(N-1)}\sin(\varphi_N). \tag{5.2c}$$

Starting from the phase at one junction boundary φ_1, the phase develops according to these recursive formulas (5.2) depending on the bias current I and the external phase $\Delta\varphi^{\text{ext}}$.

For numerical calculations with various but fixed I and $\Delta\varphi^{\text{ext}}$, we guess values of the initial phase $\varphi_1 \in [0, 2\pi)$, apply the Eqs. (5.2a) and (5.2b), and verify by Eq. (5.2c), whether the chosen initial phase conform to the boundary conditions or not. A valid solution stands out due to current conservation, which means that the injected current at the one junction boundary equals the ejected current at the other one.

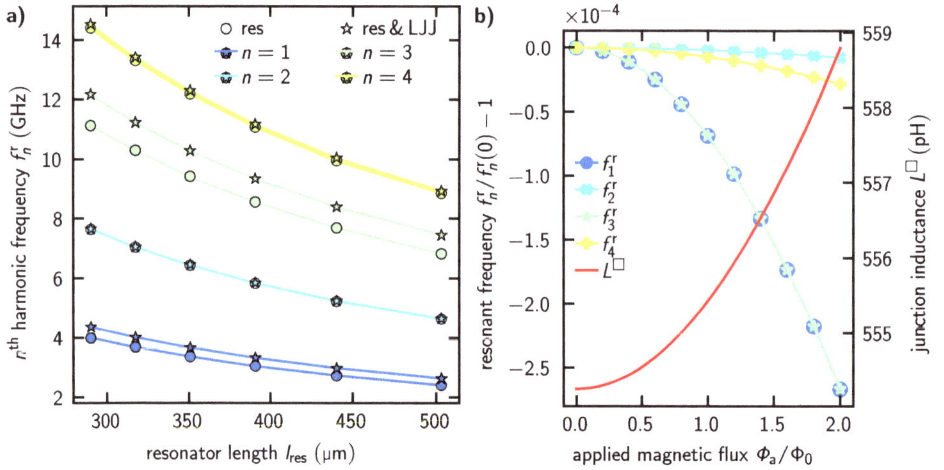

Figure 5.5: Numerically simulated anharmonicity and magnetic field dependence of a phase-biased junction. a) Odd harmonics of hybrid systems of resonators with LJJs (stars) feature significantly increased eigenfrequencies compared to reference resonators without any junction (dots). **b)** This fundamental distinction is also evident in the resonant frequencies' magnetic field dependence, whose odd harmonics are affected way stronger by the LJJ than even harmonics.

By running a nested interval algorithm, multiple mathematical solutions are found, where the energetically most favorable solution is picked as the physical solution. The phase drops along the junction $\varphi_{N-1} - \varphi_0$ for different bias currents yields a current-phase relation $I(\Delta\varphi)$, which is fitted to a fifth-degree polynomial. The coefficient of the linear term equals the reciprocal of the inductance, which is inserted in the FEM simulation to obtain the resonant frequencies.

The simulated resonant frequencies of the first four harmonics, as well as their magnetic field dependence, are depicted in Fig. 5.5. Both results show that the coupling strength between LJJ and resonator depends on the excited harmonic, and thus confirm the earlier discussed considerations.

5.4 Spectroscopic characterization

The transmission spectrum of the complex S_{21} parameter exhibits numerous resonances, each as described in Sec. 2.3.2. These are different harmonics of the frequency multiplexed resonators, which first have to be assigned to the corresponding circuits (see Fig. 5.6 a)). The fundamental modes match the results of FEM simulations and are governed by the differing resonator length l, as width, capacitance per unit length, and sheet inductance are constant.

Table 5.1: Measured and simulated anharmonicity of resonators with and without an embedded long junction in zero magnetic fields. The anharmonicities of the first three modes from spectroscopy measurements fit exactly to the *Ansys HFSS* FEM simulations in the case of the bare reference resonator with a sheet inductance of $L^{\square} = 5.649\,\mathrm{nH}$. Assuming the same inductance for the AlO_x layer of the resonator and the junction's sheet inductance of $L^{\square} = 0.554\,\mathrm{nH}$, the FEM simulations show good agreement with the measured resonance frequencies.

circuit	f_2/f_1		f_3/f_1	
	experiment	simulation	experiment	simulation
bare resonator	1.915 ± 0.005	1.915 ± 0.003	2.807 ± 0.005	2.807 ± 0.014
resonator & LJJ	1.768 ± 0.005	1.759 ± 0.004	2.76 ± 0.02	2.807 ± 0.014

The higher harmonics are attributed to the fundamental modes by two-tone spectroscopy measurements, whose schematic setup is explained in Sec. 3.2.3. Thereby, the fundamental mode is probed, while the second tone's frequency is swept with a way higher power. If the second tone is in resonance with any harmonic, the resonator is populated so much that its fundamental mode shifts to lower frequencies due to the nonlinear inductance. Using this Kerr effect and the well-known resonator geometries, the resonant frequencies can be attributed to the frequency multiplexed resonators, as shown in Fig. 5.6.

The reference resonator (#3) without any coupled LJJ can be unambiguously identified since it shows almost no magnetic field dependence at fields far below H_{c2} of AlO_x (see Fig. C.1 in Appendix C) [Bor+20] and as its harmonics match FEM simulations as listed in Tab. 5.1. From the comparison of the measured and simulated resonance frequencies, a sheet inductance of $L^{\square}_{AlO_x} = 5.469\,\mathrm{nH}$ can be estimated. This value agrees with the resistivity measurements of the test structures at room temperature and originates in the already discussed overetching of the sample.

5.4.1 Anharmonicity of the coupled system

Their relative spatial current distributions for the other five coupled systems of resonators and LJJs are identical since each LJJ covers the inner 10 % of the resonator. This similarity is why the fundamental modes of these coupled systems show a uniform length dependence, as can be seen in Fig. 5.6 a). Compared to the reference resonator, the fundamental resonant frequency is increased by about 7.3 %, which is in good agreement with the FEM simulations described in the model of phase-biased junctions in Sec. 5.3.

In contrast, both resonators with and without an embedded LJJ feature a uniform length dependence of their second harmonic frequency. This is because the low-

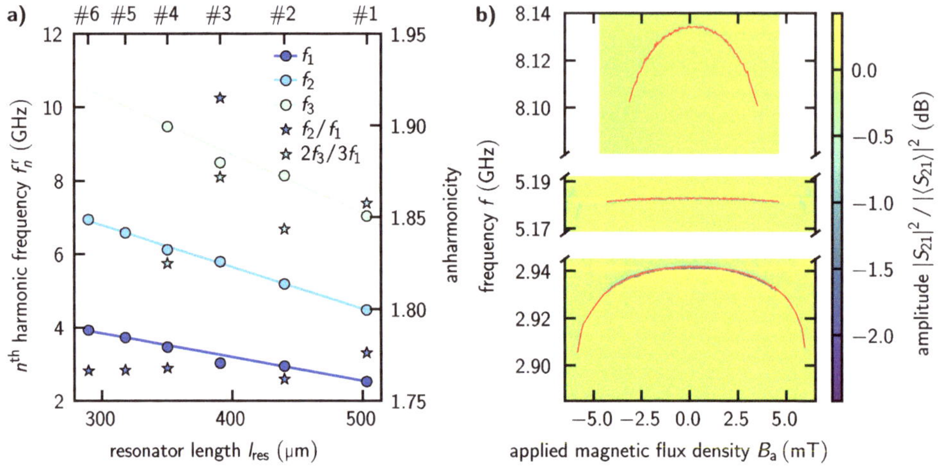

Figure 5.6: First three harmonic resonance frequencies of different resonators with and without LJJ in the zero-field and their magnetic field dependence. a) The measured resonant frequencies of the coupled systems of the resonator and long junction in zero magnetic field show a length dependence illustrated by straight lines to guide the eye. Unlike the second harmonic of the reference resonator (#3), the resonant frequency deviates from this course for odd harmonics. This different behavior of circuit #3 is apparent from the anharmonicity plotted as stars on the right ordinate. **b)** The magnetic field-dependent amplitude signal of a resonator with LJJ (exemplarily shown for circuit #2) exhibits a significantly larger shift to lower resonant frequencies for odd than for even harmonics.

inductive shunt of the junction is placed in the current node, and thus modifies the phase drop along this segment only slightly.

The third harmonics show an increased resonant frequency similar to the first harmonics because there is again a current antinode at the position of the LJJ. Unfortunately, not all third harmonics of the six circuits are measurable, but the remaining four values resemble the theory. The fourth and higher harmonics are regrettably not measurable.

A comparison of the average measured and simulated anharmonicities of the resonators with and without LJJ show good agreement between theory and experiment, as listed in Tab. 5.1. The slight deviation originates in the assumption for the simulation of the long junctions' homogeneous sheet inductance that, in reality, spatially differs. Especially at the long junctions' edge regions where the supercurrent mainly tunnels through the tunnel barrier, the inductance peaks as the vertical Josephson inductance adds to the lead inductance.

5.4.2 Magnetic field dependence of the resonant frequencies

An LJJ embedded in a resonator affects not only the anharmonicity but also the magnetic field dependence, which was theoretically discussed in Sec. 5.3. Microwave spectroscopy data confirm a comparable dependency experimentally, as exemplarily depicted for the circuit #2 in Fig. 5.6 b). Since the reference resonator does not feature such a field dependence, but is constant in its resonant frequency far below H_{c2} (see Fig. C.1 in Appendix C), it is reasonable to attribute this effect to the additional LJJ.

5.5 Concluding remarks

These proof of principle measurements show that LJJs can be galvanically coupled to microwave resonators by extending the bottom electrode to a microstrip resonator that biases the junction at microwave frequencies. The combination of LJJ position and current distribution of the excited resonator mode is crucial for the resonant frequency of the coupled system and its magnetic field dependence. Spectroscopy measurements of these two parameters show good agreement with FEM simulations, in which the phase drop along the junction is numerically calculated by the model of a phase-biased junction. These simulations further reveal that the coupling strength of the two systems varies with the inductance ratio of the bottom and top electrodes and the width ratio of the junction and resonator.

This RF manipulation and readout scheme has been used for the first time for long junctions and offers tremendous advantages for experimentally investigating quantum coherence effects in long junctions. Since the resonator is a narrowband Purcell filter and the circuit is galvanically isolated from the environment, this strong decoherence channel is significantly reduced compared to DC measurements.

6 Josephson vortices as quantum particles

In the previous chapters, we could demonstrate by transport measurements (see Sec. 4) that the junction impedance, and thus the quantumness of long junctions, can be increased with additional kinetic inductance in the junction electrodes. With the RF measurement scheme described in Sec. 5, an appropriate method was found for quantum coherence measurements. We are now ready to investigate the vortices' quantum character with these prerequisites. For this purpose, very narrow junctions are patterned from high-impedance trilayers and coupled to a microwave resonator each. These samples are characterized by steady-state microwave spectroscopy and pulsed microwave measurements.

6.1 Sample considerations and design

In principle, the same sample design is used as described in Sec. 5.1, but the LJJs are significantly narrowed to increase the inherent vortices' quantumness. A junction width of $w = 200\,\mathrm{nm}$ is a good tradeoff between this aimed quantumness and reliably producible structure sizes without creating constrictions or broadenings that would cause pinning potentials for vortices [SBM97]. Moreover, tens of microns long and 200 nm wide AlO_x strips as junction electrodes are still wide enough to ensure a certain phase of the superconducting wave function and avoid phase slips [Vos+21].

To read out the system's quantum state using the dispersive shift of the resonator, it is mandatory that the resonator's eigenfrequency remains stable over time. The previous measurements of high-impedance resonators (see Sec. 5), however, have shown fluctuations of the resonant frequency in the order of hundreds of kilohertz. Such strong variations are known from different works about superconducting films near the superconductor to insulator transition and appear if the sheet resistance exceeds the superconducting resistance quantum $R_\mathrm{q} = h/(2e)^2 = 6.45\,\mathrm{k\Omega}$ [FI18; Kri22]. In the circuits of trilayer D, the high sheet resistance $R_\mathrm{n}^\square = (4.6 \pm 0.6)\,\mathrm{k\Omega}$ results from the already large intended value of $R_\mathrm{n}^\square = 2.3\,\mathrm{k\Omega}$ and the overetching.

Therefore, to reduce frequency fluctuations in new samples, AlO_x bottom layers with a sheet resistance of $R_\mathrm{n}^\square = 1\,\mathrm{k\Omega}$ at an enlarged thickness of $d_1 = 50\,\mathrm{nm}$

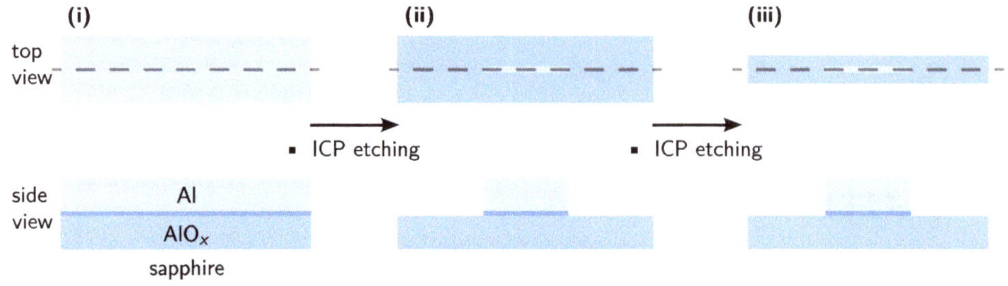

Figure 6.1: Schematic patterning of junction stacks for microwave measurements. The functional microstructure's top and side view after each fabrication step are illustrated in the upper and lower row, respectively. Starting from the deposited junction stacks **(i)**, the long junctions' top electrodes are patterned in by electron-beam lithography and an argon-chlorine-based inductively coupled plasma (ICP) etching process **(ii)**. The resonators **(iii)** are defined afterwards by optical lithography and ICP etching of the remaining bottom electrode, too.

are targeted. Since the slope of $R(d)$ (see Fig. 3.2) decreases with larger film thicknesses, the sheet resistance of thicker films is less sensitive to overetching. According to Eq. (2.28), the desired sheet resistance of $R_{\mathrm{n}}^{\square} = 1\,\mathrm{k}\Omega$ corresponds to a sheet inductance of $L_{\mathrm{k}}^{\square} = 0.75\,\mathrm{nH}$.

Again, a 30 nm thick layer of pure aluminum is used as top electrode to strongly couple junctions to resonators. Moreover, the coupling strength is varied by designing with a width of 2 µm, 4 µm, and 8 µm, while the junction width remains fixed at 200 nm. To obtain resonant frequencies in the range 4.5 GHz–7.5 GHz for the fundamental modes, a 750 µm and a 900 µm long resonator are designed for each resonator width. The junctions lengths are still $0.1 \cdot l$, i.e., 75 µm and 90 µm.

6.2 Sample fabrication details

Due to the similar design, the fabrication process is related to the one of LJJs coupled to microwave resonators, which is presented in Sec. 5.2. The main difference is the requirement of electron beam (e-beam) lithography to resolve the LJJs' structure size of 200 nm.

This process step will be performed before patterning isolated microstrip resonators to prevent damaging the LJJ by electrostatic discharge due to the deposited electrons during e-beam lithography. Due to the much smaller tunnel resistance over the extensive area of 2 cm × 2 cm, the risk of such damage is reduced significantly. The deposited charge can further be led away from the still completely connected trilayer much easier via clamps.

The photoresist *ma-N 2403* is suitable as it is sensitive to both e-beam and deep UV, it provides a resolution of $\gtrsim 50\,\text{nm}$, and it is resilient against argon-chlorine-based inductively coupled plasma (ICP) reactive ion etching (RIE). Moreover, using this negative photoresist saves exposure time since its exposed areas persist in the development while the unexposed resist is lifted so that only the tiny junction areas need to be written by the electron beam. With this method, illustrated in Fig. 6.1, long junctions are patterned from trilayer E, whose characteristic parameters are given in Tab. 3.1. The resonators are defined in the second process step, using photolithography and ICP etching as before. The subsequent deposition of a backside metallization is analog to previous samples and consists of 5 nm titanium and 100 nm silver. See Appendix B.2 for detailed parameters concerning optical and e-beam lithography (Tabs. B.1, B.2, B.3, and B.4) and ICP etching (Tab. B.5).

6.3 Energy spectrum of vortex tunneling

Spectroscopy measurements show resonances with internal and coupling quality factor $Q_\text{i} = (1.4 \pm 0.4) \times 10^5$ and $Q_\text{c} = 1828 \pm 4$ in the single photon limit, whose resonance frequencies fluctuate insignificantly. The overall magnetic field dependence resembles previous spectroscopy measurements of a LJJ embedded in a resonator, reported in Sec. 5.4, described with the model of phase-biased junctions.

In addition, pairs of avoided level crossings are observed that appear at particular flux densities $\tilde{B}_n$, as can be seen in Fig. 6.2 and Fig. 6.3 a). Such features indicate that the driven resonator exchanges energy with a coupled system at a similar bare resonant frequency. As long as the coupled system is sufficiently nonlinear, its mode spectrum vs. applied magnetic field can be tracked by two-tone spectroscopy, similar as explained in Sec. 2.3.3. For this purpose the resonance dip is determined at each field value, the resonant frequency is probed with the VNA f_p and the second tone's drive frequency f_d is swept with an additional microwave source (for the setup see Fig. 3.5 a)). If this extra drive tone excites the second mode, the resonance curve shifts dispersively so that both phase and amplitude signals change drastically. As shown in Fig. 6.2 b), such two-tone spectroscopy data exhibit hyperbolas, which are an unambiguous signature of quantum tunneling in a two-level system that is associated with a double-well potential.

The ratios of the magnetic flux density values $\tilde{B}_n$, at which these pairs of avoided level crossings occur, match the intersections of adjacent fluxons states in the magnetic diffraction pattern as shown in Fig. 6.3. Thus, the two states can be understood as $|n-1\rangle$ and $|n\rangle$ vortices in the LJJ, whose eigenenergies are similar,

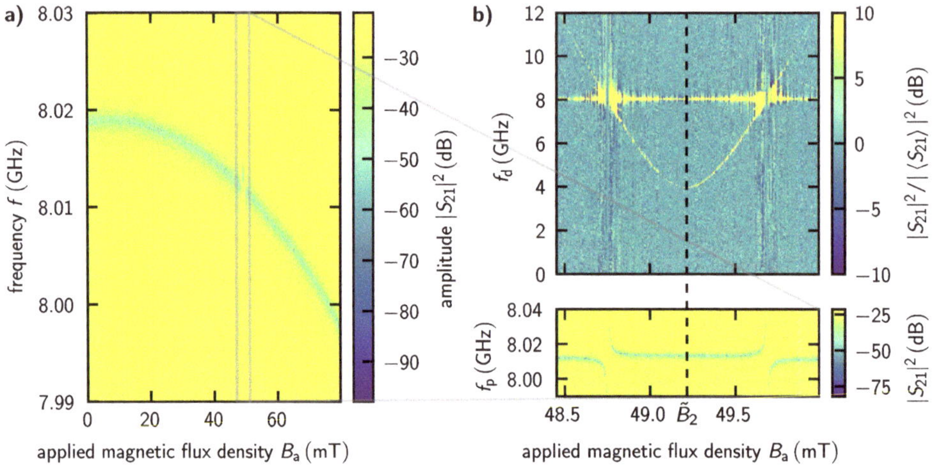

Figure 6.2: Magnetic field-dependent spectrum and two-tone spectroscopy data. a) The overall magnetic field dependence of the samples of trilayer E resembles those of the samples D as expected since the basic design is the same, and only the dimensions of the long junction differ. Therefore, the system of a long junction embedded in a resonator can further be described by the model of a phase-biased junction. In addition, sets of two mirrored avoided level crossings are observed at particular magnetic field values. **b)** Two-tone spectroscopy on one of these anticrossings shows a mode that depends hyperbolically on the applied magnetic flux density. The constant line at $f = 8.015\,\mathrm{GHz}$ can be attributed to the resonator, which naturally also changes its resonant frequency because of the self-Kerr effect when strongly driven by the second tone.

so that a single vortex can tunnel in and out of the junction. Within the scope of the standard tunneling model (STM), this tunneling between two energetically similar states is described by the Hamiltonian [FLS64b; Leg+87]

$$\mathcal{H}_{\mathrm{STM}} = \frac{1}{2}\begin{pmatrix} \varepsilon & \Delta \\ \Delta & -\varepsilon \end{pmatrix} = \frac{1}{2}\varepsilon\hat{\sigma}_z + \frac{1}{2}\Delta\hat{\sigma}_x. \tag{6.1}$$

Here, Δ denotes the coupling energy between the lowest bound states in each potential well, so to say the tunneling amplitude, ε is the energy asymmetry of the two wells, and $\hat{\sigma}_x$, $\hat{\sigma}_z$ the Pauli matrices. The double well potential is symmetric in the sweet spots $\tilde{B}_n$ where the $|n-1\rangle$ and $|n\rangle$ fluxons states are energetically equal and by overcoming the energy at the junction boundary, one fluxon can tunnel in and out of the LJJ. At applied magnetic flux densities slightly below the sweet spot $B_a \lesssim \tilde{B}_n$, the $|n-1\rangle$ state becomes more favorable, whereas the $|n\rangle$ state is more likely for slightly larger flux densities $B_a \gtrsim \tilde{B}_n$. The off-diagonal elements in the STM-Hamiltonian of Eq. (6.1) couple the $|n-1\rangle$ and $|n\rangle$ states through tunneling, so that the first two eigenstates of the two wells hybridize as

$$|\psi_-\rangle = \cos\left(\frac{\vartheta}{2}\right)|n-1\rangle - \sin\left(\frac{\vartheta}{2}\right)|n\rangle\,, \tag{6.2a}$$

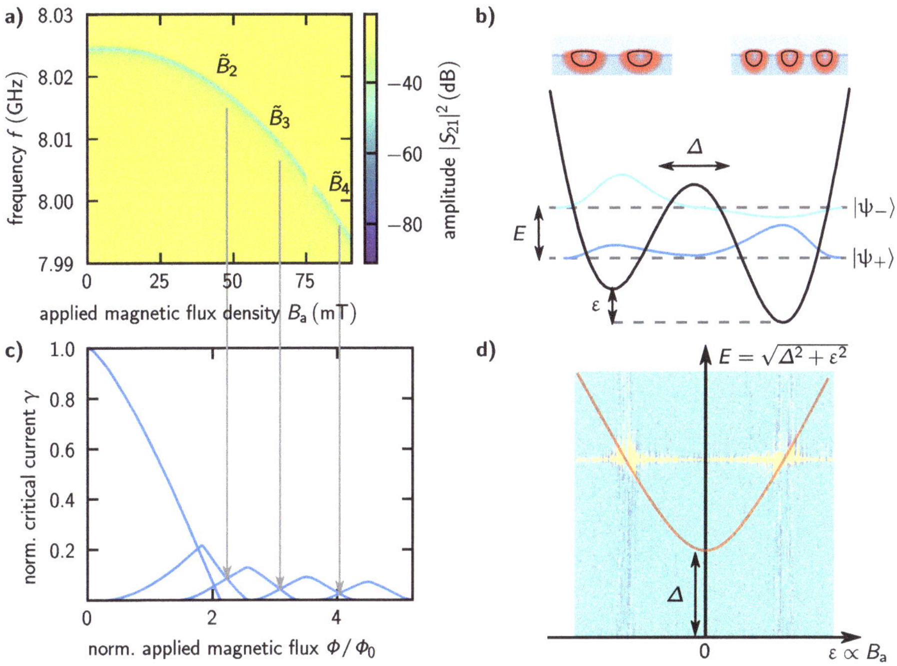

Figure 6.3: Operating principle of a Josephson vortex qubit. a) The $|S_{21}|^2$-spectrum against applied magnetic flux shows three pairs of avoided level crossings, whose magnetic flux densities $\tilde{B}_n$ coincide with the simulated intersections of two adjacent fluxon states for a normalized junction length of $\tilde{\ell} \approx 7$ **c)** (indicated by gray arrows). For these values, the states of $|n-1\rangle$ and $|n\rangle$ fluxons are energetically similar, so a single fluxon can tunnel into the junction. In doing so, the fluxon has to overcome the energy of the junction boundary, which can be modeled as a double well potential **c)**. In the sweet spot, where the two energies are identical ($E_n = E_{n-1}$), the potential is symmetric ($\varepsilon = 0$) and the fluxon is maximally delocalized between the two minima, so to say in the junction inside and outside. Off the sweet spot $|\varepsilon| \gg \Delta$ the fluxon is trapped in one of the two minima. **d)** The energy difference of the two hybridized eigenstates $|\Psi_\pm\rangle$ shows a hyperbola as a function of the asymmetry ε and can be excited as shown by two-tone spectroscopy (see Fig. 6.2 b)).

$$|\psi_+\rangle = \sin\left(\frac{\vartheta}{2}\right)|n-1\rangle + \cos\left(\frac{\vartheta}{2}\right)|n\rangle\,, \tag{6.2b}$$

where the mixing angle $\vartheta = \arctan(\Delta/\varepsilon)$ is introduced. In the basis of these eigenstates, the STM-Hamiltonian reads $\mathcal{H}_{\mathrm{STM}} = \frac{1}{2}E\hat{\sigma}_z$ with the energy difference between the eigenstates

$$E = \sqrt{\Delta^2 + \varepsilon^2}. \tag{6.3}$$

The energy of Eq. (6.3) exactly represents the previous mentioned hyperbola, measured as a function of the applied magnetic flux, and thus the asymmetry of the two potential wells (see Fig. 6.2 b) and Fig. 6.3 d)).

6.4 Coherence properties of vortex tunneling

Knowing the transition energy between the ground and excited state from two-tone spectroscopy measurements, the quantum system can be driven with pulsed microwave signals to explore its coherence properties. For these measurements, a magnetic field bias at the sweet spot is suitable since the influence of field noise on the qubit frequency is minimal there. For the following pulsed microwave measurements, a magnetic flux density of $B_\mathrm{a} \approx \tilde{B}_2 \approx 48.44\,\mathrm{mT}$ is applied, which cannot be guaranteed to permanently correspond to the exact sweet spot due to temporal instability of the magnetic bias over the measuring period of several days.

6.4.1 Rabi oscillations

In the experiment, the system is repeatedly excited from the ground state by x-pulses of variable duration τ, meaning rotated around the x-axis to different angles θ, and read out immediately afterwards. Including decoherence and energy relaxation, the projection of this rotation with constant Rabi frequency Ω_R on the z-axis results in a damped cosinusoidal oscillation

$$\langle \hat{\sigma}_z \rangle = \mathrm{e}^{-\tau/T_\mathrm{Rabi}} \cos\left(\frac{2\pi\tau}{2t_\pi}\right). \tag{6.4}$$

Figure 6.4 a) depicts such Rabi oscillations, which are characterized by alternating excitations due to photon absorption and deexcitations due to stimulated emission. From these measured data, on the one hand, the time constant of the envelope $T_\mathrm{Rabi} = 2/(T_1^{-1} + T_2^{-1})$ can be determined, which is given by the relaxation time T_1 and dephasing time T_2. On the other hand, the half-period duration t_π can be determined from the Rabi frequency, which is as the pulse length to excite the qubit from $|0\rangle$ to $|1\rangle$, i.e., to perform a π-rotation on the Bloch sphere (compare to Sec. 2.3.4).

The quantum Rabi frequency depends both on the drive frequency and amplitude, as explained in Sec. 2.3.4. For Rabi measurements with different drive frequency f_d, Chevron patterns are expected according to Eq. (2.52) (see Fig. 2.9 a) for theoretical simulations). Measured data, as exemplarily depicted in Fig. 6.4 b), are in good accordance to theoretical predictions since the Rabi frequency speeds up while the amplitude decreases with increasing detuning. These effects can be explained by the fact that the driven rotation is no longer around the pure x-axis, but around a tilted axis in the xz-plane [SN17; Bra18; Sch20]. As a consequence, the rotations still start at $|0\rangle$, but $|1\rangle$ cannot be reached anymore due to the tilt.

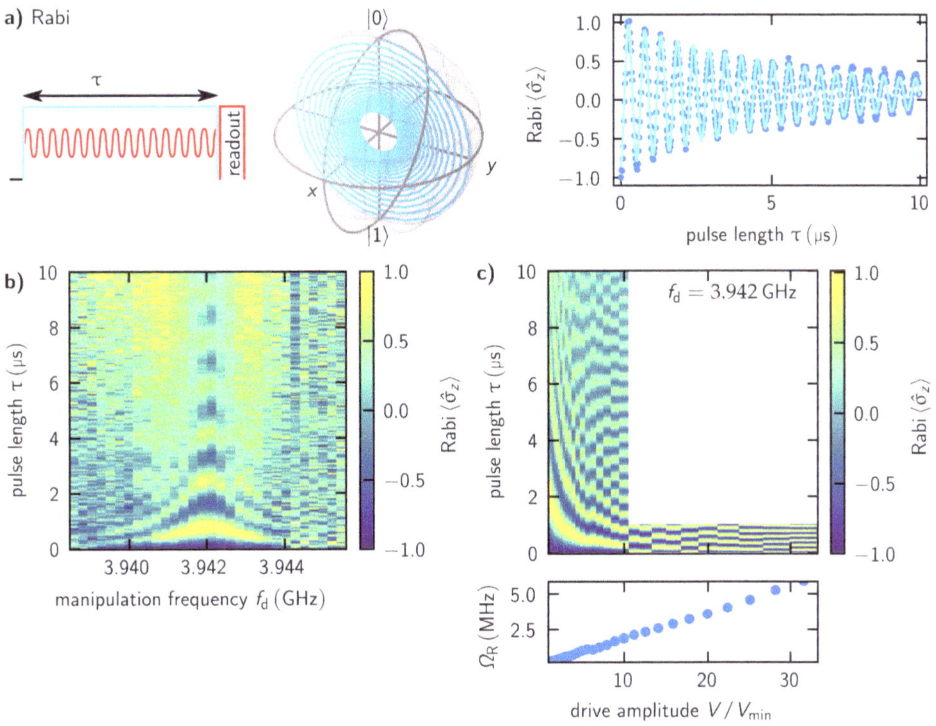

Figure 6.4: Quantum mechanical Rabi oscillations in a Josephson vortex qubit. **a)** Repeatedly applied manipulation pulses of variable duration τ rotate the qubit state by different azimuth angles on the Bloch sphere, so that the expectation value $\langle \hat{\sigma}_z \rangle$ features Rabi oscillations. **b)** The Rabi frequency depends on the drive frequency or rather the detuning from the qubit frequency. Eq. (2.52) reproduces the measured data and exhibits the typical Chevron pattern. **c)** For quantum mechanical Rabi oscillations, their frequency increases linearly with the amplitude of the drive tone. The good agreement of measured data and the theory proves the quantum nature of the Josephson vortex and confirms the device to be a Josephson vortex qubit.

In Fig. 6.4 c) the quantum Rabi oscillation's linear dependence on the drive amplitude V is shown, which is in excellent agreement with the theoretical expectations. This power dependence is characteristic of quantum mechanical Rabi oscillations, whose bare Rabi frequency without detuning equals the drive amplitude V. In other words, at higher powers, there are, on average more photons in the system, which drive the qubit more strongly, thus faster. The linear slope of the generalized Rabi frequencies Ω_R against the drive amplitude V finally proves the Josephson vortex' quantum character. Together with the hyperbolic spectrum, shown in Fig. 6.2 b), these data manifest that a Josephson vortex tunnels coherently in the system under consideration, which makes the device a Josephson vortex qubit.

6.4.2 Relaxation and dephasing times

The envelope of Rabi oscillations shows an exponential decay caused by interactions with the environment. This coupling to the environment is indispensable since qubits are intended to be manipulated and read out from outside. Two processes can be fundamentally distinguished:

First, real qubits relax from the excited state to the ground state under spontaneous emission into the resonator, the environment, or internal modes in the solids such as defects. The time constant of this exponential decay is usually called energy relaxation time T_1. In the Bloch sphere illustration, the energy relaxation manifests as a decay along the z-axis into $|0\rangle$, so that the qubit state is no longer bound to the sphere's surface but may be in its interior.

Second, fluctuations in the transition frequency ω_{01} lead to the dephasing of the qubit. Averaging multiple measurement values with random dephasing reveals a decay towards zero within the xy-plane. For example, magnetic field noise leads to a fluctuating phase, and thus to another exponential decay with the characteristic dephasing time T_2.

Relaxation time

To measure the energy relaxation time T_1, the qubit is brought into the excited state $|1\rangle$ by means of a t_π-pulse. After a delay time Δt, the probability of the excited state $\langle\hat{\sigma}_z\rangle$ is determined using the resonator's dispersive shift. During the time Δt the qubit is free to evolve and decays to the ground state, which is described by

$$\langle\hat{\sigma}_z\rangle = 1 - 2\mathrm{e}^{-\Delta t/T_1}. \tag{6.5}$$

Figure 6.5 a) depicts the schematic measurement protocol, the illustration in the Boch sphere, and the measured $\langle\hat{\sigma}_z\rangle$ data. By fitting these data to Eq. (6.5), the energy relaxation time T_1 can be determined, whose statistical distribution is shown in Fig. 6.5 d). For this purpose, the T_1 measurement is repeated 1465 times within $\approx 86\,\mathrm{h}$ and yields on average $T_1 = (45 \pm 5)\,\mu\mathrm{s}$.

Ramsey dephasing time

To determine the dephasing time T_2, a "Ramsey" measurement is performed, where the qubit state is brought with a $t_\pi/2$-pulse around the x-axis to the state $\frac{1}{\sqrt{2}}(|0\rangle + |1\rangle)$ on the equator. Arriving in the xy-plane, the qubit evolves freely for a variable duration Δt, which is nothing but an effective Lamor precession at the detuning frequency $\Delta\omega$ in the rotating frame. The quantum state can only be read

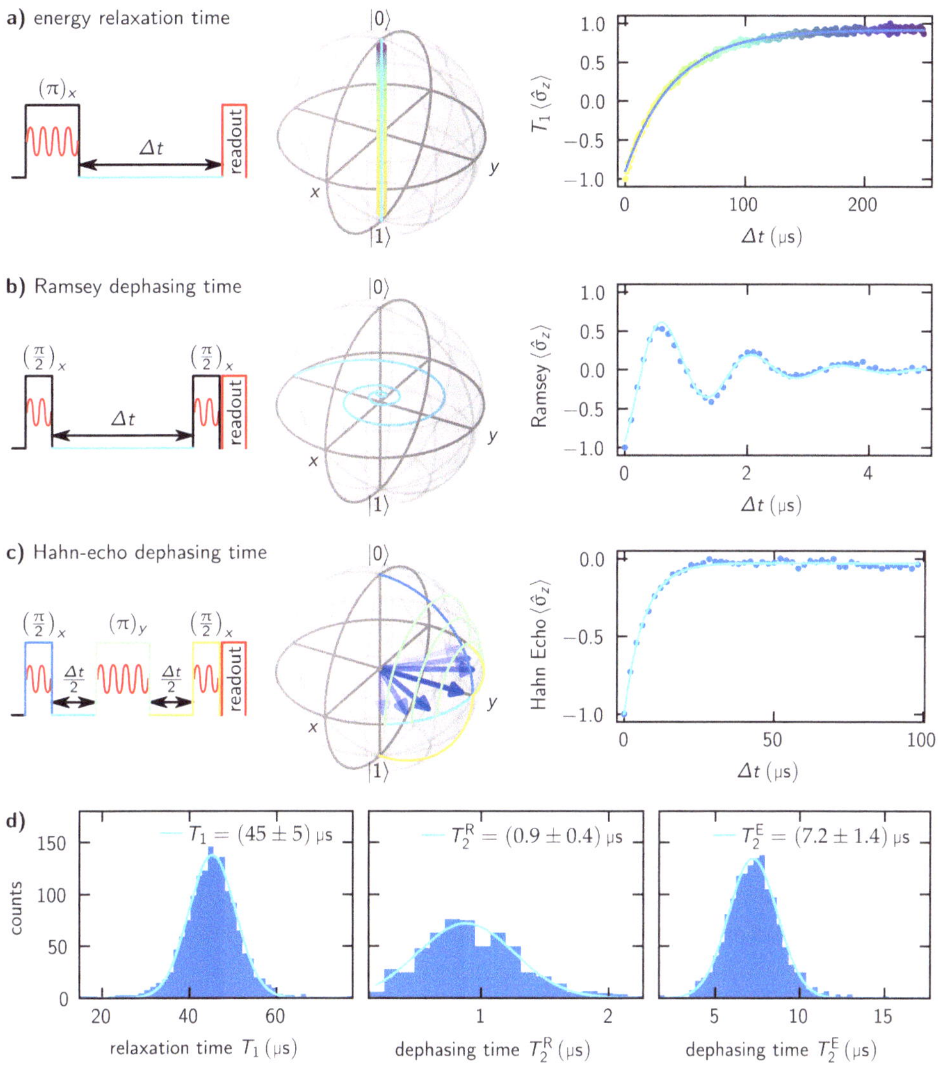

Figure 6.5: Energy relaxing and dephasing time measurements on a Josephson vortex qubit. a) To determine the energy relaxation time T_1 experimentally, the qubit is excited to $|1\rangle$ with a t_π-pulse and after a waiting time Δt the qubit state is read out. During the free evolution Δt the qubit decays exponentially along the z-axis, where the half-life is the energy relaxation time T_1. **b)** Since dephasing occurs perpendicular to z, but only the projection on z is measurable, in Ramsey dephasing measurements the qubit is rotated to and from the equator with a $t_\pi/2$-pulse at the beginning and just before the measurement, respectively. In between, the qubit evolves freely in the xy-plane for time Δt. Thereby detuning leads to an effective Lamor precession and dephasing to a decay towards the origin of the Bloch sphere, which in combination results in a damped oscillation. **c)** Low-frequency dephasing can be circumvented in Hahn-echo measurements by a t_π-pulse around y after half of the free evolution. The individual pulses are illustrated on the Bloch sphere in the same color. **d)** Long-term measurements for ≈ 86 h of T_1, T_2^R, and T_2^E yield Gaussian distributions.

out by projection on the z-axis so that a second $t_\pi/2$-pulse around the x-axis is necessary. Without dephasing (as it is the case for $\Delta t = 0$), this pulse sequence effectively corresponds to a t_π-pulse, split in two parts, so that the qubit ends up in $|1\rangle$. With dephasing, however, the state rotates around the z-axis, and thus cosinusoidal in y, which is rotated by the second $t_\pi/2$-pulse in the z-direction.[1] In addition to the effective Lamor precession due to detuning, dephasing causes the qubit to decay towards the origin of the Bloch sphere. In consequence, the overall result is a damped oscillation

$$\langle\hat{\sigma}_z\rangle = -\mathrm{e}^{-\Delta t/T_2^{\mathrm{R}}}\cos(\Delta\omega\Delta t) \tag{6.6}$$

with half-life T_2^{R}. Since for very small detunings $\Delta\omega$, the first quarter wave of the oscillation would be difficult to distinguish from the superimposed exponential decay, and thus challenging to fit, one usually deliberately chooses a detuning $\Delta\omega \sim 1\,\mathrm{MHz}$ to see a few oscillations during the decay. The Ramsey pulse sequence, its illustration on the Bloch sphere, and example measurement data are depicted in Fig. 6.5 b). Long-term measurements for $\approx 86\,\mathrm{h}$ with 568 repetitions give the distribution shown in Fig. 6.5 d) and a mean Ramsey dephasing time of $T_2^{\mathrm{R}} = (0.9 \pm 0.4)\,\mu\mathrm{s}$.

Hahn-echo dephasing time

To become insensitive to dephasing with slower frequencies than the inverse measurement duration, a t_π-pulse around y can be introduced at half of the free evolution time, i.e., after $\Delta t/2$ (see Fig. 6.5 c)). This pulse mirrors the state on the y-axis and the two rotations around the z-axis during $\Delta t/2$ partially compensate each other. In these so-called "Hahn-echo" or "spin-echo" measurements, an exponential decay

$$\langle\hat{\sigma}_z\rangle = \mathrm{e}^{-\Delta t/T_2^{\mathrm{E}}} \tag{6.7}$$

is obtained with the characteristic Hahn-echo dephasing time $T_2^{\mathrm{E}} \geq T_2^{\mathrm{R}}$ (see Fig. 6.5 c)). The significantly longer Hahn-echo dephasing time $T_2^{\mathrm{E}} = (7.2 \pm 1.4)\,\mu\mathrm{s}$ (for 1441 measurement repetitions) indicates that the Ramsey dephasing time strongly suffers from low-frequency noise of the qubit transition frequency. One major noise source is current noise of the magnet coil. Persistent current coils, for example, could provide a remedy for this.

[1] This is why, the measured $\langle\hat{\sigma}_z\rangle$ is plotted in Fig. 6.5 b), although the qubit decays in the xy-plane.

6.5 Concluding remarks

In summary, we have demonstrated the, to our knowledge, first experimental realization of a Josephson vortex qubit using a quasi-one-dimensional high-impedance long junction. At particular applied magnetic flux densities, the device features coherent quantum tunneling of a single vortex in and out of the junction, forming a two-level system. Such quantum systems have a characteristic hyperbolic mode spectrum, observed experimentally by two-tone spectroscopy and strongly indicates the increased quantumness of the inherent Josephson vortices.

Moreover, pulsed microwave measurements reveal quantum Rabi oscillations, whose frequency increases proportionally to the applied drive frequency. This observation again proves the Josephson vortex' quantum nature. The average energy relaxation time of $T_1 \approx 45\,\mu s$ is comparable to state-of-the-art superconducting qubits. This result is impressive against the background that the measured qubit is the very first of its kind, focusing on the principle implementation but not on optimized relaxation and coherence times. With that said, the average Hahn-echo dephasing time of $T_2^E \approx 7\,\mu s$ is appealing, too.

T_2 should be theoretically limited by $2T_1$, which is why much longer T_2 times can be expected with an optimized setup regarding magnetic field noise. Apart from these limitations, which could be similarly solved for other superconducting qubit types, the vortex qubits devised in this work are a promising development for the research field of superconducting quantum circuits and the long-standing dream of quantum computing.

7 Summary and outlook

The goal of this work was to experimentally investigate the quantum coherence of fluxons in quasi-one-dimensional long Josephson junctions. Entering the Josephson vortices' quantum regime is experimentally challenging to reach [Wal+03b] since such excitations in conventional junctions occur on an energy scale that exceeds the system's quantized eigenenergies by orders of magnitude. In consequence, the Josephson medium and its inherent excitations are pretty classical.

The long junction's crucial energy scale decreases with its increasing impedance $Z_\mathrm{J} = \left(L^\square / c_\mathrm{J} w^2\right)^{1/2}$ and is defined by the electrodes' sheet inductance $L^\square$, the specific junction capacitance c_J, and its width w. The approach followed in this work was to replace bulk junction electrodes with thin films of high-kinetic inductance superconductors in addition to very narrow junctions. The kinetic inductance in strips of the superconductor granular aluminum oxide exceeds the geometrical part by several orders of magnitude. It thus facilitates an increased quantumness of the long junction, as outlined in Sec. 2.4.4.

Within the scope of this work, we developed fabrication techniques for impedance-tailored junctions featuring good controllability of the kinetic inductance during sputter deposition and, if desired, high impedances.

With transport characterization measurements at millikelvin temperatures, we demonstrate high quality factors of the junctions. Long junctions also show typical magnetic diffraction patterns of their critical currents and zero-field and Fiske steps, which we use to determine the Josephson length and the Swihart velocity, respectively. These zero-field steps prove the existence of Josephson vortex excitations in the high-impedance long junctions. Furthermore, we determined the Josephson plasma frequency by switching current measurements under microwave irradiation.

The results of these three independent measurements demonstrate that the sine-Gordon model is valid without restrictions for high-impedance long junctions and $\bar{c} = \lambda_\mathrm{J}\omega_\mathrm{p}$ is still satisfied. Here, λ_J and $\bar{c}$ are reduced by about one order of magnitude due to the additional kinetic inductance, while ω_p it is nearly unaffected. The reduced Swihart velocity is accompanied by an impedance increase up to $14\,\Omega$, which corresponds to the largest value ever measured in long junctions to our knowledge [Wil+22]. From these measurement results, we conclude that the chosen approach is suitable to increase Josephson vortices' quantumness so that they could behave like quantum particles.

A suitable measurement scheme for quantum coherence properties is possible in the framework of circuit quantum electrodynamics, namely a Josephson medium connected to a microwave resonator. For this purpose, we galvanically couple high-impedance long junctions to a distributed microstrip resonator each, which we manipulate and read out by microwave photons. By measuring the anharmonicity and the magnetic field dependence of the coupled system's resonant frequency, we attribute the junction's influence on the resonator unambiguously. This result confirms the coupling of the two systems and is understood by the theoretical treatment of phase-biased junctions.

We combine our findings gathered from previous experiments and focus on the intrinsic quantumness of Josephson vortices by significantly increasing the impedance of the long junction through high kinetic electrode inductances and narrow junction widths. This junction is embedded in a microwave resonator for manipulation and readout with single photons. In microwave spectroscopy measurements versus applied magnetic flux density, we observe hyperbolic modes at particular fields. This mode spectrum is characteristic of quantum tunneling in a two-level system. We attribute this signature to a tunneling process of a single fluxon in and out of the long junction since measured magnetic field values correspond to the intersections between $|n-1\rangle$ and $|n\rangle$ fluxon states from theoretical predictions.

By driving the transition of this two-level system by pulsed microwave signals, we observe quantum Rabi oscillations, whose frequency is proportional to the drive amplitude. This measurement demonstrates the quantum nature of the vortices for the first time to our knowledge. The vortex qubit features a measured mean energy relaxation time of $T_1 = (45 \pm 5)\,\mu s$ and an average Hahn-echo dephasing time of $T_2^{H} = (7.2 \pm 1.4)\,\mu s$.

The experimental methods developed in this work could also be used in various future experiments. For instance, the coupling of the Josephson junction to a microwave resonator suggests using these circuits to study the internal mode spectrum of the long junction and to directly excite, e.g., the Josephson plasma frequency, zero-field, and Fiske resonances.

Besides, the high-kinetic inductance electrodes enable impedance-tailored junctions to match external circuits and $50\,\Omega$ cables. This impedance matching is crucial for increasing the efficiency of Josephson flux-flow oscillators applied for microwave generation and amplification.

The discovery of Josephson quantum vortices is a promising starting point for further experimental and theoretical work. One could study the dynamic properties of such quantum particles in the nonlinear Josephson medium, i.e., quantum solitons within the scope of the quantum sine-Gordon model [Roy+21].

Beyond these uniform media, one could design more sophisticated junction geometries, such as spatially periodic structures, including potential barriers and wells. They can be realized by broadenings ("microshorts") or constrictions ("microresistors") of the long junction, where the quantum vortices are repelled or pinned. Quantum coherent entities in such metamaterials should lead to a band structure for propagating collective linear and nonlinear excitations, similar to phonons in crystals.

The last point, we want to outline here, is optimizing the properties of the Josephson vortex qubit. The priority should be to reduce its dephasing by filtering the magnetic field noise or, ideally, using persistent current coils. Furthermore, alternative vortex qubits are also possible. For instance, microshort- and mircoresistor-impurities can be used to design two-level systems whose pinning and repulsion potentials allow for coherent tunneling.

Bibliography

[ACC66] B. Abeles, R. W. Cohen, and G. W. Cullen: *Enhancement of Superconductivity in Metal Films*. Physical Review Letters **17** (1966), 632–634. DOI: 10.1103/PhysRevLett.17.632 (cit. on p. 56).

[Aff81] I. Affleck: *Quantum-Statistical Metastability*. Physical Review Letters **46** (1981), 388–391. DOI: 10.1103/PhysRevLett.46.388 (cit. on p. 75).

[AP95] G. L. Alfimov and A. F. Popkov: *Magnetic vortices in a distributed Josephson junction with electrodes of finite thickness*. Physical Review B **52** (1995), 4503–4510. DOI: 10.1103/PhysRevB.52.4503 (cit. on p. 40).

[AB63] V. Ambegaokar and A. Baratoff: *Tunneling Between Superconductors*. Physical Review Letters **10** (1963), 486–489. DOI: 10.1103/PhysRevLett.10.486 (cit. on pp. 31, 76, 79).

[AR63] P. W. Anderson and J. M. Rowell: *Probable Observation of the Josephson Superconducting Tunneling Effect*. Physical Review Letters **10** (1963), 230–232. DOI: 10.1103/PhysRevLett.10.230 (cit. on pp. 1, 30).

[ASB89] S. M. Anlage, H. J. Snortland, and M. R. Beasley: *A current controlled variable delay superconducting transmission line*. IEEE Transactions on Magnetics **25** (1989), 1388–1391. DOI: 10.1109/20.92554 (cit. on pp. 16, 17).

[Ann+10] A. J. Annunziata, D. F. Santavicca, L. Frunzio, G. Catelani, M. J. Rooks, A. Frydman, and D. E. Prober: *Tunable superconducting nanoinductors*. Nanotechnology **21** (2010), 445202. DOI: 10.1088/0957-4484/21/44/445202 (cit. on p. 56).

[Ann10] A. J. Annunziata: *Single-Photon Detection, Kinetic Inductance, and Non-Equilibrium Dynamics in Niobium and Niobium Nitride Superconducting Nanowires*. PhD thesis. Yale University, 2010. URL: https://proberlab.yale.edu/sites/default/files/files/Annunziata_Thesis.pdf (cit. on p. 16).

[ADK18] P. W. Atkins, J. De Paula, and J. Keeler: *Physical chemistry*. 11th ed. Oxford, New York: Oxford University Press, 2018. URL: https://academia.edu/51098021 (cit. on p. 62).

[Bar50] J. Bardeen: *Zero-Point Vibrations and Superconductivity*. Physical Review **79** (1950), 167–168. DOI: 10.1103/PhysRev.79.167.3 (cit. on p. 12).

[Bar+10] R. Barends, N. Vercruyssen, A. Endo, P. J. de Visser, T. Zijlstra, T. M. Klapwijk, and J. J. A. Baselmans: *Reduced frequency noise in superconducting resonators*. Applied Physics Letters **97** (2010), 033507. DOI: 10.1063/1.3467052 (cit. on p. 17).

[BJV75] A. Barone, W. J. Johnson, and R. Vaglio: *Current flow in large Josephson junctions*. Journal of Applied Physics **46** (1975), 3628–3632. DOI: 10.1063/1.322089 (cit. on pp. 42, 44, 45, 78).

[BP82] A. Barone and G. Paterno: *Physics and Applications of the Josephson Effect*. John Wiley & Sons, Ltd, 1982. DOI: 10.1002/352760278X (cit. on pp. 1, 31, 32, 37, 44, 47, 76, 82, 136).

[BL64] C. P. Bean and J. D. Livingston: *Surface Barrier in Type-II Superconductors*. Physical Review Letters **12** (1964), 14–16. DOI: 10.1103/PhysRevLett.12.14 (cit. on p. 1).

[BK03] K. H. Bennemann and J. B. Ketterson, eds.: *The Physics of Superconductors*. Vol. 1. 2 vols. Berlin, Heidelberg: Springer, 2003. DOI: 10.1007/978-3-642-55675-3 (cit. on p. 12).

[BK04] K. H. Bennemann and J. B. Ketterson, eds.: *The Physics of Superconductors*. Vol. 2. 2 vols. Berlin, Heidelberg: Springer, 2004. DOI: 10.1007/978-3-642-18914-2 (cit. on p. 12).

[Bes06] M. Bestehorn: *Hydrodynamik und Strukturbildung*. Springer Berlin, Heidelberg, 2006. 392 pp. DOI: 10.1007/3-540-33797-0 (cit. on p. 1).

[BCG10] J. A. Blackburn, M. Cirillo, and N. Grønbech-Jensen: *On the classical model for microwave induced escape from a Josephson washboard potential*. Physics Letters A **374** (2010), 2827–2830. DOI: 10.1016/j.physleta.2010.04.077 (cit. on p. 80).

[Bla+04] A. Blais, R.-S. Huang, A. Wallraff, S. M. Girvin, and R. J. Schoelkopf: *Cavity quantum electrodynamics for superconducting electrical circuits: An architecture for quantum computation*. Physical Review A **69** (2004), 062320. DOI: 10.1103/PhysRevA.69.062320 (cit. on pp. 18, 29, 85).

[Blo46] F. Bloch: *Nuclear Induction*. Physical Review **70** (1946), 460–474. DOI: 10.1103/PhysRev.70.460 (cit. on p. 27).

[BTŠ58] N. N. Bogoljubov, V. V. Tolmachov, and D. V. Širkov: *A New Method in the Theory of Superconductivity*. Fortschritte der Physik **6** (1958), 605–682. DOI: 10.1002/prop.19580061102 (cit. on p. 13).

[Bol+08] A. T. Bollinger, R. C. Dinsmore, A. Rogachev, and A. Bezryadin: *Determination of the Superconductor-Insulator Phase Diagram for One-Dimensional Wires*. Physical Review Letters **101** (2008), 227003. DOI: 10.1103/PhysRevLett.101.227003 (cit. on p. 17).

[Bor+20] K. Borisov, D. Rieger, P. Winkel, F. Henriques, F. Valenti, et al.: *Superconducting granular aluminum resonators resilient to magnetic fields up to 1 Tesla*. Applied Physics Letters **117** (2020), 120502. DOI: 10.1063/5.0018012 (cit. on pp. 55, 93).

[Bra18] J. Braumüller: *Quantum simulation experiments with superconducting circuits*. 2018. DOI: 10.5445/KSP/1000081315 (cit. on pp. 19, 21, 28, 102).

[BH54] W. Buckel and R. Hilsch: *Einfluß der Kondensation bei tiefen Temperaturen auf den elektrischen Widerstand und die Supraleitung für verschiedene Metalle*. Zeitschrift für Physik **138** (1954), 109–120. DOI: 10.1007/BF01337903 (cit. on p. 56).

[BHL83] M. Büttiker, E. P. Harris, and R. Landauer: *Thermal activation in extremely underdamped Josephson-junction circuits*. Physical Review B **28** (1983), 1268–1275. DOI: 10.1103/PhysRevB.28.1268 (cit. on p. 75).

[CF91] J. G. Caputo and N. Flytzanis: *Kink-antikink collisions in sine-Gordon and $ \varphi^4 $ models: Problems in the variational approach*. Physical Review A **44** (1991), 6219–6225. DOI: 10.1103/PhysRevA.44.6219 (cit. on pp. 47, 136).

[CFL71] J. T. Chen, T. F. Finnegan, and D. N. Langenberg: *Anomalous dc current singularities in Josephson tunnel junctions*. Physica **55** (1971), 413–420. DOI: 10.1016/0031-8914(71)90282-5 (cit. on pp. 1, 82).

[CHS71] P. V. Christiansen, E. B. Hansen, and C. J. Sjöström: *Negative self-inductance in superconducting thin wires and weak links*. Journal of Low Temperature Physics **4** (1971), 349–389. DOI: 10.1007/BF00628738 (cit. on p. 16).

[Chu+81a] T. Chui, G. Deutscher, P. Lindenfeld, and W. L. McLean: *Conduction in granular aluminum near the metal-insulator transition*. Physical Review B **23** (1981), 6172–6175. DOI: 10.1103/PhysRevB.23.6172 (cit. on p. 56).

[Chu+81b] T. Chui, P. Lindenfeld, W. L. McLean, and K. Mui: *Coupling and isolation: Critical field and transition temperature of superconducting granular aluminum*. Physical Review B **24** (1981), 6728–6731. DOI: 10.1103/PhysRevB.24.6728 (cit. on p. 55).

[Cir+97] M. Cirillo, T. Doderer, S. G. Lachenmann, F. Santucci, and N. Grønbech-Jensen: *Dynamical evidence of critical fields in Josephson junctions*. Physical Review B **56** (1997), 11889–11896. DOI: 10.1103/PhysRevB.56.11889 (cit. on p. 83).

[Cir+98] M. Cirillo, N. Grønbech-Jensen, M. R. Samuelsen, M. Salerno, and G. V. Rinati: *Fiske modes and Eck steps in long Josephson junctions: Theory and experiments*. Physical Review B **58** (1998), 12377–12384. DOI: 10.1103/PhysRevB.58.12377 (cit. on p. 83).

[Cla66] J. Clarke: *A superconducting galvanometer employing Josephson tunnelling*. The Philosophical Magazine: A Journal of Theoretical Experimental and Applied Physics **13** (1966), 115–127. DOI: 10.1080/14786436608211991 (cit. on p. 30).

[CW08] J. Clarke and F. K. Wilhelm: *Superconducting quantum bits*. Nature **453** (2008), 1031–1042. DOI: 10.1038/nature07128 (cit. on p. 30).

[CA68] R. W. Cohen and B. Abeles: *Superconductivity in Granular Aluminum Films*. Physical Review **168** (1968), 444–450. DOI: 10.1103/PhysRev.168.444 (cit. on pp. 55, 58).

[Col15] P. Coleman: *Introduction to Many-Body Physics*. Cambridge: Cambridge University Press, 2015. DOI: 10.1017/CBO9781139020916 (cit. on p. 13).

[Czy17] G. Czycholl: *Theoretische Festkörperphysik Band 2: Anwendungen: Nichtgleichgewicht, Verhalten in äußeren Feldern, kollektive Phänomene*. 2017. DOI: 10.1007/978-3-662-53701-5 (cit. on pp. 12, 13).

[Dah+68] A. J. Dahm, A. Denenstein, T. F. Finnegan, D. N. Langenberg, and D. J. Scalapino: *Study of the Josephson Plasma Resonance*. Physical Review Letters **20** (1968), 859–863. DOI: 10.1103/PhysRevLett.20.859 (cit. on pp. 34, 79).

[Dau+21] H. Dausy, L. Nulens, B. Raes, M. J. Van Bael, and J. Van de Vondel: *Impact of Kinetic Inductance on the Critical-Current Oscillations of Nanobridge SQUIDs*. Physical Review Applied **16** (2021), 024013. DOI: 10.1103/PhysRevApplied.16.024013 (cit. on p. 17).

[DP19] P. G. De Gennes and P. A. Pincus: *Superconductivity of Metals and Alloys*. Boca Raton: CRC Press, 2019. 292 pp. DOI: 10.1201/9780429497032 (cit. on pp. 11, 12).

[DF61] B. S. Deaver and W. M. Fairbank: *Experimental Evidence for Quantized Flux in Superconducting Cylinders*. Physical Review Letters **7** (1961), 43–46. DOI: 10.1103/PhysRevLett.7.43 (cit. on pp. 1, 8).

[DM08] D. Depla and S. Mahieu, eds.: *Reactive Sputter Deposition*. Red. by R. Hull, R. M. Osgood, J. Parisi, and H. Warlimont. Vol. 109. Springer Series in Materials Science. Berlin, Heidelberg: Springer, 2008. DOI: 10.1007/978-3-540-76664-3 (cit. on p. 57).

[Deu+73a] G. Deutscher, H. Fenichel, M. Gershenson, E. Grünbaum, and Z. Ovadyahu: *Transition to zero dimensionality in granular aluminum superconducting films*. Journal of Low Temperature Physics **10** (1973), 231–243. DOI: 10.1007/BF00655256 (cit. on p. 56).

[Deu+73b] G. Deutscher, M. Gershenson, E. Grünbaum, and Y. Imry: *Granular Superconducting Films*. Journal of Vacuum Science and Technology **10** (1973), 697–701. DOI: 10.1116/1.1318416 (cit. on p. 56).

[DMC85] M. H. Devoret, J. M. Martinis, and J. Clarke: *Measurements of Macroscopic Quantum Tunneling out of the Zero-Voltage State of a Current-Biased Josephson Junction*. Physical Review Letters **55** (1985), 1908–1911. DOI: 10.1103/PhysRevLett.55.1908 (cit. on pp. 74, 75).

[DN61] R. Doll and M. Näbauer: *Experimental Proof of Magnetic Flux Quantization in a Superconducting Ring*. Physical Review Letters **7** (1961), 51–52. DOI: 10.1103/PhysRevLett.7.51 (cit. on pp. 1, 8).

[Due+81] B. Dueholm, E. Joergensen, O. A. Levring, J. Mygind, N. F. Pedersen, M. R. Samuelsen, O. H. Olsen, and M. Cirillo: *Dynamic fluxon model for Fiske steps in long Josephson junctions*. Physica B+C **108** (1981), 1303–1305. DOI: 10.1016/0378-4363(81)90951-7 (cit. on p. 83).

[EGK15] E. G. Ekomasov, A. M. Gumerov, and R. V. Kudryavtsev: *On the possibility of the observation of the resonance interaction between kinks of the sine-Gordon equation and localized waves in real physical systems*. JETP Letters **101** (2015), 835–839. DOI: 10.1134/S0021364015120061 (cit. on p. 47).

[EH05] C. Enss and S. Hunklinger: *Low-Temperature Physics*. Berlin, Heidelberg: Springer, 2005. DOI: 10.1007/b137878 (cit. on pp. 1, 62).

[EFP83] S. N. Erné, A. Ferrigno, and R. D. Parmentier: *Fluxon propagation and Fiske steps in long Josephson tunnel junctions*. Physical Review B **27** (1983), 5440–5446. DOI: 10.1103/PhysRevB.27.5440 (cit. on p. 83).

[FI18] M. V. Feigel'man and L. B. Ioffe: *Microwave Properties of Superconductors Close to the Superconductor-Insulator Transition*. Physical Review Letters **120** (2018), 037004. DOI: 10.1103/PhysRevLett.120.037004 (cit. on p. 97).

[FP63] R. A. Ferrell and R. E. Prange: *Self-Field Limiting of Josephson Tunneling of Superconducting Electron Pairs*. Physical Review Letters **10** (1963), 479–481. DOI: 10.1103/PhysRevLett.10.479 (cit. on pp. 38, 39, 78).

[FLS64a] R. P. Feynman, R. B. Leighton, and M. L. Sands: *The Feynman Lectures on Physics: Electromagnetism and Matter*. Vol. 2. Addison-Wesley Pub. Co., 1964. URL: https://www.feynmanlectures.caltech.edu/II_toc.html (cit. on p. 30).

[FLS64b] R. P. Feynman, R. B. Leighton, and M. L. Sands: *The Feynman Lectures on Physics: Quantum mechanics*. Vol. 3. Addison-Wesley Pub. Co., 1964. URL: https://www.feynmanlectures.caltech.edu/III_toc.html (cit. on p. 100).

[FFT73] B. F. Field, T. F. Finnegan, and J. Toots: *Volt Maintenance at NBS via 2e/h: A New Definition of the NBS Volt*. Metrologia **9** (1973), 155–166. DOI: 10.1088/0026-1394/9/4/003 (cit. on p. 30).

[Fis64] M. D. Fiske: *Temperature and Magnetic Field Dependences of the Josephson Tunneling Current*. Reviews of Modern Physics **36** (1964), 221–222. DOI: 10.1103/RevModPhys.36.221 (cit. on pp. 1, 83).

[FU03] M. V. Fistul and A. V. Ustinov: *Josephson vortex interaction mediated by cavity modes: Tunable coupling for superconducting qubits*. Physical Review B **68** (2003), 132509. DOI: 10.1103/PhysRevB.68.132509 (cit. on p. 2).

[Fri44] H. T. Friis: *Noise Figures of Radio Receivers*. Proceedings of the IRE **32** (1944), 419–422. DOI: 10.1109/JRPROC.1944.232049 (cit. on p. 64).

[Frö50] H. Fröhlich: *Theory of the Superconducting State. I. The Ground State at the Absolute Zero of Temperature*. Physical Review **79** (1950), 845–856. DOI: 10.1103/PhysRev.79.845 (cit. on p. 12).

[FD74a] T. A. Fulton and L. N. Dunkleberger: *Lifetime of the zero-voltage state in Josephson tunnel junctions*. Physical Review B **9** (1974), 4760–4768. DOI: 10.1103/PhysRevB.9.4760 (cit. on p. 73).

[FD74b] T. A. Fulton and L. N. Dunkleberger: *Vortex propagation and radiation emission in Josephson tunnel junctions*. Rev.Phys.Appl. **9** (1974), 299–303. DOI: 10.1051/rphysap:0197400901029900 (cit. on pp. 1, 82).

[FD73] T. A. Fulton and R. C. Dynes: *Single vortex propagation in Josephson tunnel junctions*. Solid State Communications **12** (1973), 57–61. DOI: 10.1016/0038-1098(73)90345-1 (cit. on pp. 1, 82).

[GC08] J. Garrison and R. Chiao: *Quantum Optics*. Oxford Graduate Texts. Oxford: Oxford University Press, 2008. 730 pp. DOI: 10.1093/acprof:oso/9780198508861.001.0001 (cit. on p. 25).

[Geb22] R. Gebauer: *A Flexible FPGA-based Control Platform for Superconducting Multi-Qubit Experiments*. PhD thesis. 2022. DOI: 10.5445/IR/1000141695 (cit. on p. 68).

[Geb+20a] R. Gebauer, N. Karcher, D. Gusenkova, M. Spiecker, L. Grünhaupt, et al.: *State preparation of a fluxonium qubit with feedback from a custom FPGA-based platform*. AIP Conference Proceedings **2241** (2020), 020015. DOI: 10.1063/5.0011721 (cit. on p. 68).

[Geb+20b] R. Gebauer, N. Karcher, J. Hurst, M. Weber, and O. Sander: *Accelerating complex control schemes on a heterogeneous MPSoC platform for quantum computing* (2020). DOI: 10.48550/arXiv.2004.07755 (cit. on p. 68).

[Gia+06] F. Giazotto, T. T. Heikkilä, A. Luukanen, A. M. Savin, and J. P. Pekola: *Opportunities for mesoscopics in thermometry and refrigeration: Physics and applications*. Reviews of Modern Physics **78** (2006), 217–274. DOI: 10.1103/RevModPhys.78.217 (cit. on p. 75).

[GL09] V. L. Ginzburg and L. D. Landau: "On the Theory of Superconductivity". *On Superconductivity and Superfluidity: A Scientific Autobiography*. Ed. by V. L. Ginzburg. Berlin, Heidelberg: Springer, 2009, pp. 113–137. DOI: 10.1007/978-3-540-68008-6_4 (cit. on p. 8).

[GT57] R. E. Glover and M. Tinkham: *Conductivity of Superconducting Films for Photon Energies between 0.3 and $ 40kT_c $*. Physical Review **108** (1957), 243–256. DOI: 10.1103/PhysRev.108.243 (cit. on p. 56).

[GPS14] H. Goldstein, C. P. Poole, and J. L. Safko: *Classical Mechanics*. 3rd ed. Pearson, 2014. URL: https://pearson.com/en-us/subject-catalog/p/classical-mechanics/P200000006871/9780201657029 (cit. on p. 48).

[Gor59] L. P. Gor'Kov: *Microscopic derivation of the Ginzburg-Landau equations in the theory of superconductivity*. Soviet Physics JETP **36 (9)** (1959), 1364–1367. URL: http://www.jetp.ras.ru/cgi-bin/dn/e_09_06_1364.pdf (cit. on p. 10).

[GC34] C. J. Gorter and H. Casimir: *On supraconductivity I*. Physica **1** (1934), 306–320. DOI: 10.1016/S0031-8914(34)90037-9 (cit. on pp. 9, 77).

[Gra+08] G. J. Grabovskij, L. J. Swenson, O. Buisson, C. Hoffmann, A. Monfardini, and J.-C. Villégier: *In situ measurement of the permittivity of helium using microwave NbN resonators*. Applied Physics Letters **93** (2008), 134102. DOI: 10.1063/1.2996263 (cit. on p. 17).

[Grø+04] N. Grønbech-Jensen, M. G. Castellano, F. Chiarello, M. Cirillo, C. Cosmelli, L. V. Filippenko, R. Russo, and G. Torrioli: *Microwave-Induced Thermal Escape in Josephson Junctions*. Physical Review Letters **93** (2004), 107002. DOI: 10.1103/PhysRevLett.93.107002 (cit. on p. 80).

[GMD16] R. Gross, A. Marx, and F. Deppe: *Applied Superconductivity: Josephson Effect and Superconducting Electronics*. De Gruyter Textbook Series. Walter De Gruyter Incorporated, 2016 (cit. on p. 32).

[Grü+18] L. Grünhaupt, N. Maleeva, S. T. Skacel, M. Calvo, F. Levy-Bertrand, A. V. Ustinov, H. Rotzinger, A. Monfardini, G. Catelani, and I. M. Pop: *Loss Mechanisms and Quasiparticle Dynamics in Superconducting Microwave Resonators Made of Thin-Film Granular Aluminum*. Physical Review Letters **121** (2018), 117001. DOI: 10.1103/PhysRevLett.121.117001 (cit. on p. 55).

[Grü+19] L. Grünhaupt, M. Spiecker, D. Gusenkova, N. Maleeva, S. T. Skacel, et al.: *Granular aluminium as a superconducting material for high-impedance quantum circuits*. Nature Materials **18** (2019), 816–819. DOI: 10.1038/s41563-019-0350-3 (cit. on p. 55).

[Gu+17] X. Gu, A. F. Kockum, A. Miranowicz, Y.-x. Liu, and F. Nori: *Microwave photonics with superconducting quantum circuits*. Physics Reports. Microwave photonics with superconducting quantum circuits **718-719** (2017), 1–102. DOI: 10.1016/j.physrep.2017.10.002 (cit. on p. 18).

[Ham00] C. A. Hamilton: *Josephson voltage standards*. Review of Scientific Instruments **71** (2000), 3611–3623. DOI: 10.1063/1.1289507 (cit. on p. 30).

[HR06] S. Haroche and J.-M. Raimond: *Exploring the Quantum: Atoms, Cavities, and Photons*. Oxford Graduate Texts. Oxford: Oxford University Press, 2006. 616 pp. DOI: 10.1093/acprof:oso/9780198509141.001.0001 (cit. on p. 29).

[Haz+19] T. M. Hazard, A. Gyenis, A. Di Paolo, A. T. Asfaw, S. A. Lyon, A. Blais, and A. A. Houck: *Nanowire Superinductance Fluxonium Qubit*. Physical Review Letters **122** (2019), 010504. DOI: 10.1103/PhysRevLett.122.010504 (cit. on p. 17).

[HSB94] Z. Hermon, A. Stern, and E. Ben-Jacob: *Quantum dynamics of a fluxon in a long circular Josephson junction*. Physical Review B **49** (1994), 9757–9762. DOI: 10.1103/PhysRevB.49.9757 (cit. on p. 52).

[IS90] Y. M. Ivanchenko and T. K. Soboleva: *Nonlocal interaction in Josephson junctions*. Physics Letters A **147** (1990), 65–69. DOI: 10.1016/0375-9601(90)90015-G (cit. on p. 40).

[Jak+64] R. C. Jaklevic, J. Lambe, A. H. Silver, and J. E. Mercereau: *Quantum Interference Effects in Josephson Tunneling*. Physical Review Letters **12** (1964), 159–160. DOI: 10.1103/PhysRevLett.12.159 (cit. on p. 30).

[JC63] E. T. Jaynes and F. W. Cummings: *Comparison of quantum and semiclassical radiation theories with application to the beam maser*. Proceedings of the IEEE **51** (1963), 89–109. DOI: 10.1109/PROC.1963.1664 (cit. on p. 29).

[JV09] D. Jeltsema and A. J. Van Der Schaft: *Lagrangian and Hamiltonian formulation of transmission line systems with boundary energy flow*. Reports on Mathematical Physics **63** (2009), 55–74. DOI: 10.1016/S0034-4877(09)00009-3 (cit. on p. 47).

[Jos62] B. D. Josephson: *Possible new effects in superconductive tunnelling*. Physics Letters **1** (1962), 251–253. DOI: 10.1016/0031-9163(62)91369-0 (cit. on pp. 1, 30).

[Jos64] B. D. Josephson: *Coupled Superconductors*. Reviews of Modern Physics **36** (1964), 216–220. DOI: 10.1103/RevModPhys.36.216 (cit. on pp. 1, 32, 38, 41, 56).

[Jos65] B. D. Josephson: *Supercurrents through barriers*. Advances in Physics **14** (1965), 419–451. DOI: 10.1080/00018736500101091 (cit. on pp. 1, 32, 47).

[Kam08] H. Kamerlingh Onnes: *The liquefaction of helium* (1908), 168. URL: https://www.dwc.knaw.nl/DL/publications/PU00013525.pdf (cit. on p. 5).

[Kam11] H. Kamerlingh Onnes: *The resistance of pure mercury at helium temperatures*. Commun. Phys. Lab. Univ. Leiden, b **120** (1911). URL: https://physics.ucf.edu/~rep/EDII/Onnes1911.pdf (cit. on p. 5).

[KI96] T. Kato and M. Imada: *Macroscopic Quantum Tunneling of a Fluxon in a Long Josephson Junction*. Journal of the Physical Society of Japan **65** (1996), 2963–2975. DOI: 10.1143/JPSJ.65.2963 (cit. on pp. 2, 52).

[Kem06] A. Kemp: *Quantum and thermal phase escape in extended Josephson systems* (2006). URL: https://www.osti.gov/etdeweb/biblio/20865471 (cit. on p. 53).

[KWU02] A. Kemp, A. Wallraff, and A. V. Ustinov: *Josephson Vortex Qubit: Design, Preparation and Read-Out*. physica status solidi (b) **233** (2002), 472–481. DOI: 10.1002/1521-3951(200210)233:3<472::AID-PSSB472>3.0.CO;2-J (cit. on pp. 2, 52).

[Kha+12] M. S. Khalil, M. J. A. Stoutimore, F. C. Wellstood, and K. D. Osborn: *An analysis method for asymmetric resonator transmission applied to superconducting devices*. Journal of Applied Physics **111** (2012), 054510. DOI: 10.1063/1.3692073 (cit. on p. 22).

[KD11] J. H. Kim and R. P. Dhungana: *Effects of a resonant cavity on macroscopic quantum tunneling of fluxons in long Josephson junctions*. Physical Review B **83** (2011), 064503. DOI: 10.1103/PhysRevB.83.064503 (cit. on p. 2).

[KDP06] J. H. Kim, R. P. Dhungana, and K.-S. Park: *Decoherence in Josephson vortex quantum bits: Long-Josephson-junction approach to a two-state system*. Physical Review B **73** (2006), 214506. DOI: 10.1103/PhysRevB.73.214506 (cit. on p. 2).

[Kit05] C. Kittel: *Introduction to Solid State Physics*. 8th ed. New York: Wiley, 2005. URL: https://www.wiley.com/en-us/Introduction+to+Solid+State+Physics+8th+Edition-p-9780471415268 (cit. on pp. 5, 14).

[KW19] R. Kleiner and H. Wang: "Intrinsic Josephson Junctions in High Temperature Superconductors". *Fundamentals and Frontiers of the Josephson Effect*. Ed. by F. Tafuri. Springer Series in Materials Science. Cham: Springer International Publishing, 2019, pp. 367–454. DOI: 10.1007/978-3-030-20726-7_10 (cit. on pp. 15, 37).

[KMM95] A. W. Kleinsasser, R. E. Miller, and W. H. Mallison: *Dependence of critical current density on oxygen exposure in Nb-AlO/sub x/-Nb tunnel junctions*. IEEE Transactions on Applied Superconductivity **5** (1995), 26–30. DOI: 10.1109/77.384565 (cit. on pp. 59, 141).

[Kos+01] V. P. Koshelets, P. N. Dmitriev, A. B. Ermakov, A. S. Sobolev, A. M. Baryshev, P. R. Wesselius, and J. Mygind: *Radiation linewidth of flux-flow oscillators*. Superconductor Science and Technology **14** (2001), 1040–1043. DOI: 10.1088/0953-2048/14/12/312 (cit. on p. 1).

[Kra+19] P. Krantz, M. Kjaergaard, F. Yan, T. P. Orlando, S. Gustavsson, and W. D. Oliver: *A quantum engineer's guide to superconducting qubits*. Applied Physics Reviews **6** (2019), 021318. DOI: 10.1063/1.5089550 (cit. on pp. 2, 27).

[Kri+19] S. Krinner, S. Storz, P. Kurpiers, P. Magnard, J. Heinsoo, R. Keller, J. Lütolf, C. Eichler, and A. Wallraff: *Engineering cryogenic setups for 100-qubit scale superconducting circuit systems*. EPJ Quantum Technology **6** (2019), 1–29. DOI: 10.1140/epjqt/s40507-019-0072-0 (cit. on p. 64).

[Kri22] M. Kristen. In preparation (2022) (cit. on p. 97).

[Kul65] I. O. Kulik: *Theory of "Steps" of Voltage-current Characteristic of the Josephson Tunnel Current*. ZhETF Pisma Redaktsiiu **2** (1965), 134. URL: http://jetpletters.ru/ps/1597/article_24482.shtml (cit. on pp. 1, 83).

[Kul67] I. O. Kulik: *Theory of the tunnel effect in a superconductor*. Soviet Physics Journal **10** (1967), 12–14. DOI: 10.1007/BF00819974 (cit. on pp. 1, 83).

[Lan37] L. D. Landau: *On the theory of phase transitions. I*. Phys. Z. Sowjet. **11** (1937). DOI: 10.1016/B978-0-08-010586-4.50034-1 (cit. on p. 8).

[LG50] L. D. Landau and V. L. Ginzburg: *On the theory of superconductivity*. Zh. Eksp. Teor. Fiz.; (USSR) **20:12** (1950), 1064–1082. DOI: 10.1016/B978-0-08-010586-4.50035-3 (cit. on p. 8).

[Lau+95] A. Laub, T. Doderer, S. G. Lachenmann, R. P. Huebener, and V. A. Oboznov: *Lorentz Contraction of Flux Quanta Observed in Experiments with Annular Josephson Tunnel Junctions*. Physical Review Letters **75** (1995), 1372–1375. DOI: 10.1103/PhysRevLett.75.1372 (cit. on p. 1).

[LS67] P. Lebwohl and M. J. Stephen: *Properties of Vortex Lines in Superconducting Barriers*. Physical Review **163** (1967), 376–379. DOI: 10.1103/PhysRev.163.376 (cit. on p. 39).

[Led+10] H. G. Leduc, B. Bumble, P. K. Day, B. H. Eom, J. Gao, et al.: *Titanium nitride films for ultrasensitive microresonator detectors*. Applied Physics Letters **97** (2010), 102509. DOI: 10.1063/1.3480420 (cit. on p. 17).

[Leg+87] A. J. Leggett, S. Chakravarty, A. T. Dorsey, M. P. A. Fisher, A. Garg, and W. Zwerger: *Dynamics of the dissipative two-state system*. Reviews of Modern Physics **59** (1987), 1–85. DOI: 10.1103/RevModPhys.59.1 (cit. on p. 100).

[Leg87] O. Legrand: *Kink-antikink dissociation and annihilation: A collective-coordinate description*. Physical Review A **36** (1987), 5068–5073. DOI: 10.1103/PhysRevA.36.5068 (cit. on pp. 47, 136).

[LPS83] O. A. Levring, N. F. Pedersen, and M. R. Samuelsen: *Perturbation calculation of magnetic field dependence of fluxon dynamics in long inline and overlap Josephson junctions*. Journal of Applied Physics **54** (1983), 987–991. DOI: 10.1063/1.332025 (cit. on p. 48).

[Lik86] K. K. Likharev: *Dynamics of Josephson junctions and circuits*. Philadelphia, Pa.: Gordon and Breach Science Publishers, 1986. URL: https://www.taylorfrancis.com/books/mono/10.1201/9781315141572/ (cit. on pp. 1, 39, 81, 82).

[Lon48] F. London: *On the Problem of the Molecular Theory of Superconductivity*. Physical Review **74** (1948), 562–573. DOI: 10.1103/PhysRev.74.562 (cit. on pp. 7, 10, 13).

[Lon50] F. London: *Macroscopic Theory of Superconductivty*. John Wiley & Sons. Superfluids **1** (1950). URL: http://w2agz.com/Library/Classic%20Papers%20in%20Superconductivity/London,%20Superfluids%20(Wiley,%20New%20York,%201950).pdf (cit. on pp. 1, 8).

[LL35] F. London and H. London: *Supraleitung und Diamagnetismus*. Physica **2** (1935), 341–354. DOI: 10.1016/S0031-8914(35)90097-0 (cit. on p. 6).

[LU08] A. Lukashenko and A. V. Ustinov: *Improved powder filters for qubit measurements*. Review of Scientific Instruments **79** (2008), 014701. DOI: 10.1063/1.2827515 (cit. on p. 65).

[Luo+14] J. Luomahaara, V. Vesterinen, L. Grönberg, and J. Hassel: *Kinetic inductance magnetometer*. Nature Communications **5** (2014), 4872. DOI: 10.1038/ncomms5872 (cit. on p. 17).

[Mae+95] M. Maezawa, M. Aoyagi, H. Nakagawa, I. Kurosawa, and S. Takada: *Specific capacitance of Nb/AlOx/Nb Josephson junctions with critical current densities in the range of 0.1–18 kA/cm²*. Applied Physics Letters **66** (1995), 2134–2136. DOI: 10.1063/1.113927 (cit. on p. 81).

[MW18] W. Martienssen and H. Warlimont: *Springer handbook of condensed matter and materials data*. 2nd ed. Springer Handbooks. Cham: Springer International Publishing, 2018. DOI: 10.1007/978-3-319-69743-7 (cit. on p. 11).

[MK83] A. Matsuda and T. Kawakami: *Fluxon Propagation on a Josephson Transmission Line*. Physical Review Letters **51** (1983), 694–697. DOI: 10.1103/PhysRevLett.51.694 (cit. on p. 37).

[MB58] D. C. Mattis and J. Bardeen: *Theory of the Anomalous Skin Effect in Normal and Superconducting Metals*. Physical Review **111** (1958), 412–417. DOI: 10.1103/PhysRev.111.412 (cit. on pp. 18, 56).

[Max50] E. Maxwell: *Isotope Effect in the Superconductivity of Mercury*. Physical Review **78** (1950), 477–477. DOI: 10.1103/PhysRev.78.477 (cit. on p. 12).

[MSJ69] A. F. Mayadas, M. Shatzkes, and J. F. Janak: *Electrical resistivity model for polycrystalline films: the case of specular reflection at external surfaces*. Applied Physics Letters **14** (1969), 345–347. DOI: 10.1063/1.1652680 (cit. on pp. 58, 59).

[McC68] D. E. McCumber: *Effect of ac Impedance on dc Voltage-Current Characteristics of Superconductor Weak-Link Junctions*. Journal of Applied Physics **39** (1968), 3113–3118. DOI: 10.1063/1.1656743 (cit. on pp. 32, 33).

[MS78] D. W. McLaughlin and A. C. Scott: *Perturbation analysis of fluxon dynamics*. Physical Review A **18** (1978), 1652–1680. DOI: 10.1103/PhysRevA.18.1652 (cit. on pp. 1, 48, 51, 52, 82).

[MO33] W. Meissner and R. Ochsenfeld: *Ein neuer Effekt bei Eintritt der Supraleitfähigkeit*. Naturwissenschaften **21** (1933), 787–788. DOI: 10.1007/BF01504252 (cit. on p. 6).

[Mer70] J. E. Mercereau: *Superconducting magnetometers*. Revue de Physique Appliquée **5** (1970), 13–20. DOI: 10.1051/rphysap:019700050101300 (cit. on p. 30).

[MT69] R. Meservey and P. M. Tedrow: *Measurements of the Kinetic Inductance of Superconducting Linear Structures*. Journal of Applied Physics **40** (1969), 2028–2034. DOI: 10.1063/1.1657905 (cit. on p. 16).

[Mon+13] R. Monaco, V. P. Koshelets, A. Mukhortova, and J. Mygind: *Self-field effects in window-type Josephson tunnel junctions*. Superconductor Science and Technology **26** (2013), 055021. DOI: 10.1088/0953-2048/26/5/055021 (cit. on pp. 42, 78).

[Nag+83] T. Nagatsuma, K. Enpuku, F. Irie, and K. Yoshida: *Flux-flow type Josephson oscillator for millimeter and submillimeter wave region*. Journal of Applied Physics **54** (1983), 3302–3309. DOI: 10.1063/1.332443 (cit. on p. 1).

[Nag+85] T. Nagatsuma, K. Enpuku, H. Iwakura, and K. Yoshida: *Flux-Flow-Type Josephson Linear Amplifier with Large Gain and Wide Linear Range*. Japanese Journal of Applied Physics **24** (1985), L599. DOI: 10.1143/JJAP.24.L599 (cit. on p. 1).

[NPT99] Y. Nakamura, Y. A. Pashkin, and J. S. Tsai: *Coherent control of macroscopic quantum states in a single-Cooper-pair box*. Nature **398** (1999), 786–788. DOI: 10.1038/19718 (cit. on p. 30).

[NSS60] J. Nicol, S. Shapiro, and P. H. Smith: *Direct Measurement of the Superconducting Energy Gap*. Physical Review Letters **5** (1960), 461–464. DOI: 10.1103/PhysRevLett.5.461 (cit. on pp. 15, 76).

[NBB19] D. Niepce, J. Burnett, and J. Bylander: *High Kinetic Inductance NbN Nanowire Superinductors*. Physical Review Applied **11** (2019), 044014. DOI: 10.1103/PhysRevApplied.11.044014 (cit. on p. 17).

[Nol14] W. Nolting: *Grundkurs Theoretische Physik 2*. Springer-Lehrbuch. Berlin, Heidelberg: Springer, 2014. DOI: 10.1007/978-3-642-41980-5 (cit. on p. 48).

[Nol18] W. Nolting: *Grundkurs Theoretische Physik 1*. Ed. by W. Nolting. Berlin, Heidelberg: Springer, 2018. DOI: 10.1007/978-3-662-57584-0 (cit. on p. 25).

[Nor95] J. E. Nordman: *Superconductive amplifying devices using fluxon dynamics*. Superconductor Science and Technology **8** (1995), 681–699. DOI: 10.1088/0953-2048/8/9/001 (cit. on p. 1).

[Ols+86] O. H. Olsen, N. F. Pedersen, M. R. Samuelsen, H. Svensmark, and D. Welner: *Perturbation treatment of boundary conditions for fluxon motion in long Josephson junctions*. Physical Review B **33** (1986), 168–173. DOI: 10.1103/PhysRevB.33.168 (cit. on p. 48).

[OD91] T. P. Orlando and K. A. Delin: *Foundations of applied superconductivity*. Addison-Wesley Pub. Co., 1991 (cit. on pp. 32, 50).

[OS67] C. S. Owen and D. J. Scalapino: *Vortex Structure and Critical Currents in Josephson Junctions*. Physical Review **164** (1967), 538–544. DOI: 10.1103/PhysRev.164.538 (cit. on pp. 39, 42–44).

[PRS91] S. Pagano, B. Ruggiero, and E. Sarnelli: *Magnetic-field dependence of the critical current in long Josephson junctions*. Physical Review B **43** (1991), 5364–5369. DOI: 10.1103/PhysRevB.43.5364 (cit. on pp. 42, 43).

[Par78] R. D. Parmentier: "Fluxons in Long Josephson Junctions". *Solutions in Action*. Ed. by K. Lonngren and A. Scott. Academic Press, 1978, pp. 173–199. DOI: 10.1016/B978-0-12-455580-8.50014-7 (cit. on p. 1).

[Ped83] N. F. Pedersen: "Solitons in Long Josephson Junctions". *Advances in Superconductivity*. Ed. by B. Deaver and J. Ruvalds. NATO Advanced Science Institutes Series. Boston, MA: Springer US, 1983, pp. 149–181. DOI: 10.1007/978-1-4613-9954-4_5 (cit. on p. 1).

[Ped86] N. F. Pedersen: "Chapter 9 - Solitons in Josephson Transmission Lines". *Modern Problems in Condensed Matter Sciences*. Ed. by S. E. Trullinger, V. E. Zakharov, and V. L. Pokrovsky. Vol. 17. Solitons. Elsevier, 1986, pp. 469–501. DOI: 10.1016/B978-0-444-87002-5.50015-1 (cit. on p. 81).

[PS73] N. F. Pedersen and K. Saermark: *Analytical solution for a Josephson-Junction model with capacitance*. Physica **69** (1973), 572–578. DOI: 10.1016/0031-8914(73)90089-X (cit. on p. 37).

[Poo+14] C. Poole, H. Farach, R. Creswick, and R. Prozorov: *Superconductivity*. 3rd ed. Elsevier Insights. Elsevier Science, 2014. DOI: 10.1016/C2012-0-07073-1 (cit. on p. 36).

[Poz11] D. M. Pozar: *Microwave Engineering*. 4th. Hoboken: Wiley, 2011. URL: https://www.wiley.com/en-us/Microwave+Engineering+4th+Edition-p-9780470631553 (cit. on p. 19).

[Pro+15] S. Probst, F. B. Song, P. A. Bushev, A. V. Ustinov, and M. P. Weides: *Efficient and robust analysis of complex scattering data under noise in microwave resonators*. Review of Scientific Instruments **86** (2015), 024706. DOI: 10.1063/1.4907935 (cit. on pp. 22, 23).

[qki] qkitgroup: *Qkit – a quantum measurement and evaluation suite in python*. URL: https://github.com/qkitgroup/qkit (cit. on p. 66).

[Rey+50] C. A. Reynolds, B. Serin, W. H. Wright, and L. B. Nesbitt: *Superconductivity of Isotopes of Mercury*. Physical Review **78** (1950), 487–487. DOI: 10.1103/PhysRev.78.487 (cit. on p. 12).

[Rot+16] H. Rotzinger, S. T. Skacel, M. Pfirrmann, J. N. Voss, J. Münzberg, S. Probst, P. A. Bushev, M. P. Weides, A. V. Ustinov, and J. E. Mooij: *Aluminium-oxide wires for superconducting high kinetic inductance circuits*. Superconductor Science and Technology **30** (2016), 025002. DOI: 10.1088/0953-2048/30/2/025002 (cit. on pp. 17, 55–57).

[Roy+21] A. Roy, D. Schuricht, J. Hauschild, F. Pollmann, and H. Saleur: *The quantum sine-Gordon model with quantum circuits*. Nuclear Physics B **968** (2021), 115445. DOI: 10.1016/j.nuclphysb.2021.115445 (cit. on pp. 2, 111).

[SN17] J. J. Sakurai and J. Napolitano: *Modern quantum mechanics*. 2nd ed. Cambridge University Press, 2017. DOI: 10.1017/9781108499996 (cit. on pp. 28, 29, 102).

[Sam+16] N. Samkharadze, A. Bruno, P. Scarlino, G. Zheng, D. P. DiVincenzo, L. DiCarlo, and L. M. K. Vandersypen: *High-Kinetic-Inductance Superconducting Nanowire Resonators for Circuit QED in a Magnetic Field*. Physical Review Applied **5** (2016), 044004. DOI: 10.1103/PhysRevApplied.5.044004 (cit. on p. 17).

[Sar+15] B. Sarabi, A. N. Ramanayaka, A. L. Burin, F. C. Wellstood, and K. D. Osborn: *Cavity quantum electrodynamics using a near-resonance two-level system: Emergence of the Glauber state*. Applied Physics Letters **106** (2015), 172601. DOI: 10.1063/1.4918775 (cit. on p. 26).

[Sar+91] E. Sarnelli, S. Pagano, B. Ruggiero, and M. Russo: *Magnetic field dependence of the critical current in long 'quasi-overlap' Josephson junctions*. IEEE Transactions on Magnetics **27** (1991), 2716–2719. DOI: 10.1109/20.133773 (cit. on p. 44).

[SMU97] V. V. Schmidt, P. Müller, and A. V. Ustinov: *The physics of superconductors*. Berlin, Heidelberg: Springer, 1997. DOI: 10.1007/978-3-662-03501-6 (cit. on p. 16).

[Sch20] A. Schneider: *Quantum Sensing Experiments with Superconducting Qubits*. 2020. DOI: 10.5445/KSP/1000118743 (cit. on pp. 22, 24, 27, 28, 30, 64, 102).

[Sch+20] Y. Schön, J. N. Voss, M. Wildermuth, A. Schneider, S. T. Skacel, M. P. Weides, J. H. Cole, H. Rotzinger, and A. V. Ustinov: *Rabi oscillations in a superconducting nanowire circuit*. npj Quantum Materials **5** (2020), 1–5. DOI: 10.1038/s41535-020-0220-x (cit. on p. 55).

[SB07] J. R. Schrieffer and J. S. Brooks: *Handbook of High-Temperature Superconductivity*. New York, NY: Springer, 2007. DOI: 10.1007/978-0-387-68734-6 (cit. on p. 12).

[Sch19] J. R. Schrieffer: *Theory of Superconductivity*. Boca Raton: CRC Press, 2019. 352 pp. DOI: 10.1201/9780429495700 (cit. on pp. 12, 13).

[Sch70] K. Schwidtal: *Type-I and Type-II Superconductivity in Wide Josephson Junctions*. Physical Review B **2** (1970), 2526–2532. DOI: 10.1103/PhysRevB.2.2526 (cit. on pp. 42, 44, 78).

[Sco69] A. C. Scott: *A Nonlinear Klein-Gordon Equation*. American Journal of Physics **37** (1969), 52–61. DOI: 10.1119/1.1975404 (cit. on p. 1).

[Sco70] W. C. Scott: *Hysteresis in the DC switching characteristics of Josephson junctions*. Applied Physics Letters **17** (1970), 166–169. DOI: 10.1063/1.1653350 (cit. on p. 37).

[SK05] P. D. Shaju and V. C. Kuriakose: *Double-well potential for a Josephson vortex qubit*. Physica C: Superconductivity and its Applications **424** (2005), 125–132. DOI: 10.1016/j.physc.2005.05.003 (cit. on p. 2).

[She+18] A. Shearrow, G. Koolstra, S. J. Whiteley, N. Earnest, P. S. Barry, F. J. Heremans, D. D. Awschalom, E. Shirokoff, and D. I. Schuster: *Atomic layer deposition of titanium nitride for quantum circuits*. Applied Physics Letters **113** (2018), 212601. DOI: 10.1063/1.5053461 (cit. on p. 17).

[SBM97] A. Shnirman, E. Ben-Jacob, and B. Malomed: *Tunneling and resonant tunneling of fluxons in a long Josephson junction*. Physical Review B **56** (1997), 14677–14685. DOI: 10.1103/PhysRevB.56.14677 (cit. on pp. 52, 97).

[Ste68] W. C. Stewart: *Current-voltage characteristics of josephson junctions*. Applied Physics Letters **12** (1968), 277–280. DOI: 10.1063/1.1651991 (cit. on pp. 32, 33).

[Swe+13] L. J. Swenson, P. K. Day, B. H. Eom, H. G. Leduc, N. Llombart, C. M. McKenney, O. Noroozian, and J. Zmuidzinas: *Operation of a titanium nitride superconducting microresonator detector in the nonlinear regime*. Journal of Applied Physics **113** (2013), 104501. DOI: 10.1063/1.4794808 (cit. on pp. 17, 23).

[Swi61] J. C. Swihart: *Field Solution for a Thin-Film Superconducting Strip Transmission Line*. Journal of Applied Physics **32** (1961), 461–469. DOI: 10.1063/1.1736025 (cit. on pp. 1, 39).

[Tay+67] B. N. Taylor, W. H. Parker, D. N. Langenberg, and A. Denenstein: *On the Use of the AC Josephson Effect to Maintain Standards of Electromotive Force*. Metrologia **3** (1967), 89–98. DOI: 10.1088/0026-1394/3/4/001 (cit. on p. 30).

[Tie+21] E. Tiesinga, P. J. Mohr, D. B. Newell, and B. N. Taylor: *CODATA recommended values of the fundamental physical constants: 2018*. Reviews of Modern Physics **93** (2021), 025010. DOI: 10.1103/RevModPhys.93.025010 (cit. on p. 8).

[Tin04] M. Tinkham: *Introduction to superconductivity*. 2nd ed. Dover books on physics. Mineola, New York: Dover Publications, 2004. URL: https://store.doverpublications.com/0486435032.html (cit. on pp. 1, 7, 11–14, 16–18, 33, 41).

[Ust98] A. V. Ustinov: *Solitons in Josephson junctions.* Physica D: Nonlinear Phenomena. Annual International Conference of the Center for Nonlinear Studies **123** (1998), 315–329. DOI: 10.1016/S0167-2789(98)00131-6 (cit. on pp. 1, 50).

[Ust05] A. V. Ustinov: *Solitons in Josephson Junctions: Physics of Magnetic Fluxons in Superconducting Junctions and Arrays.* John Wiley & Sons, 2005 (cit. on pp. 82, 83).

[Val58] J. G. Valatin: *Comments on the theory of superconductivity.* Il Nuovo Cimento (1955-1965) **7** (1958), 843–857. DOI: 10.1007/BF02745589 (cit. on p. 13).

[Vis+10] M. R. Vissers, J. Gao, D. S. Wisbey, D. A. Hite, C. C. Tsuei, A. D. Corcoles, M. Steffen, and D. P. Pappas: *Low loss superconducting titanium nitride coplanar waveguide resonators.* Applied Physics Letters **97** (2010), 232509. DOI: 10.1063/1.3517252 (cit. on p. 17).

[VD17] U. Vool and M. Devoret: *Introduction to quantum electromagnetic circuits.* International Journal of Circuit Theory and Applications **45** (2017), 897–934. DOI: 10.1002/cta.2359 (cit. on pp. 2, 45).

[Vos+21] J. N. Voss, Y. Schön, M. Wildermuth, D. Dorer, J. H. Cole, H. Rotzinger, and A. V. Ustinov: *Eliminating Quantum Phase Slips in Superconducting Nanowires.* ACS Nano **15** (2021), 4108–4114. DOI: 10.1021/acsnano.0c08721 (cit. on pp. 53, 97).

[Wal+00a] C. H. van der Wal, A. C. J. ter Haar, F. K. Wilhelm, R. N. Schouten, C. J. P. M. Harmans, T. P. Orlando, S. Lloyd, and J. E. Mooij: *Quantum Superposition of Macroscopic Persistent-Current States.* Science **290** (2000), 773–777. DOI: 10.1126/science.290.5492.773 (cit. on p. 30).

[Wal76] J. R. Waldram: *The Josephson effects in weakly coupled superconductors.* Reports on Progress in Physics **39** (1976), 751–827. DOI: 10.1088/0034-4885/39/8/002 (cit. on pp. 31, 32).

[Wal+03a] A. Wallraff, T. Duty, A. Lukashenko, and A. V. Ustinov: *Multiphoton Transitions between Energy Levels in a Current-Biased Josephson Tunnel Junction.* Physical Review Letters **90** (2003), 037003. DOI: 10.1103/PhysRevLett.90.037003 (cit. on p. 80).

[Wal+00b] A. Wallraff, Y. Koval, M. Levitchev, M. V. Fistul, and A. V. Ustinov: *Annular Long Josephson Junctions in a Magnetic Field: Engineering and Probing the Fluxon Interaction Potential.* Journal of Low Temperature Physics **118** (2000), 543–553. DOI: 10.1023/A:1004674908169 (cit. on p. 2).

[Wal+03b] A. Wallraff, A. Lukashenko, J. Lisenfeld, A. Kemp, M. V. Fistul, Y. Koval, and A. V. Ustinov: *Quantum dynamics of a single vortex*. Nature **425** (2003), 155–158. DOI: 10.1038/nature01826 (cit. on pp. 2, 84, 109).

[Wal+04] A. Wallraff, D. I. Schuster, A. Blais, L. Frunzio, R.-S. Huang, J. Majer, S. Kumar, S. M. Girvin, and R. J. Schoelkopf: *Strong coupling of a single photon to a superconducting qubit using circuit quantum electrodynamics*. Nature **431** (2004), 162–167. DOI: 10.1038/nature02851 (cit. on pp. 18, 85).

[Wal01] A. Wallraff: *Fluxon dynamics in annular Josephson junctions: from relativistic strings to quantum particles* (2001) (cit. on pp. 53, 74).

[WM08] D. F. Walls and G. J. Milburn: *Quantum optics*. Ed. by D. F. Walls and G. J. Milburn. Berlin, Heidelberg: Springer, 2008. DOI: 10.1007/978-3-540-28574-8 (cit. on p. 29).

[Wei69] M. Weihnacht: *Influence of Film Thickness on D. C. Josephson Current*. physica status solidi (b) **32** (1969), K169–K172. DOI: 10.1002/pssb.19690320259 (cit. on pp. 40, 79).

[Wil+22] M. Wildermuth, L. Powalla, J. N. Voss, Y. Schön, A. Schneider, M. V. Fistul, H. Rotzinger, and A. V. Ustinov: *Fluxons in high-impedance long Josephson junctions*. Applied Physics Letters **120** (2022), 112601. DOI: 10.1063/5.0082197 (cit. on pp. 41, 58, 59, 69, 72, 74, 78–80, 82, 84, 109).

[Wol11] E. L. Wolf: *Principles of Electron Tunneling Spectroscopy: Second Edition*. OUP Oxford, 2011. 618 pp. DOI: 10.1093/acprof:oso/9780199589494.001.0001 (cit. on p. 31).

[ZHB78] P. Ziemann, G. Heim, and W. Buckel: *Oxygen content and oxide barrier thickness in granular aluminum films*. Solid State Communications **27** (1978), 1131–1135. DOI: 10.1016/0038-1098(78)91127-4 (cit. on p. 56).

[ZB22] J. Zierep and K. Bühler: *Principles of Fluid Mechanics*. 1st ed. Springer Wiesbaden, 2022. 237 pp. DOI: 10.1007/978-3-658-34812-0 (cit. on p. 1).

[ZTH70] J. E. Zimmerman, P. Thiene, and J. T. Harding: *Design and Operation of Stable rf-Biased Superconducting Point-Contact Quantum Devices, and a Note on the Properties of Perfectly Clean Metal Contacts*. Journal of Applied Physics **41** (1970), 1572–1580. DOI: 10.1063/1.1659074 (cit. on p. 30).

List of Publications

[Wil+22] M. Wildermuth, L. Powalla, J. N. Voss, Y. Schön, A. Schneider, M. V. Fistul, H. Rotzinger, and A. V. Ustinov: *Fluxons in high-impedance long Josephson junctions*. Applied Physics Letters **120** (2022), 112601. DOI: 10.1063/5.0082197.

[Sch+20] Y. Schön, J. N. Voss, M. Wildermuth, A. Schneider, S. T. Skacel, M. P. Weides, J. H. Cole, H. Rotzinger, and A. V. Ustinov: *Rabi oscillations in a superconducting nanowire circuit*. npj Quantum Materials **5** (2020), 1–5. DOI: 10.1038/s41535-020-0220-x.

[Vos+21] J. N. Voss, Y. Schön, M. Wildermuth, D. Dorer, J. H. Cole, H. Rotzinger, and A. V. Ustinov: *Eliminating Quantum Phase Slips in Superconducting Nanowires*. ACS Nano **15** (2021), 4108–4114. DOI: 10.1021/acsnano.0c08721.

Appendix

A Circuit quantization of long Josephson junctions

A.1 sine-Gordon model of long Josephson junctions including high-kinetic inductance

Long Josephson junctions can be modeled by $N \to \infty$ infinitesimally small lumped elements of resistively and capacitively shunted junctions in z-direction, which are extended along the x-axis, and thus connected via inductive bottom and top electrodes (see Fig. A.1 a)). Using common circuit quantization techniques with N node fluxes in the top and bottom electrodes Φ_j^{t} and Φ_j^{b}, the Lagrangian holds

$$\begin{aligned}\mathcal{L} = &\sum_{j=0}^{N-1} \frac{c_{\mathrm{J}} w \Delta x}{2} \left(\partial_t \Phi_j^{\mathrm{t}} - \partial_t \Phi_j^{\mathrm{b}}\right)^2 - \sum_{j=0}^{N-1} \frac{\Phi_0 j_{\mathrm{c}} w \Delta x}{2\pi} \left(1 - \cos\left(\frac{2\pi}{\Phi_0}\left(\Phi_j^{\mathrm{t}} - \Phi_j^{\mathrm{b}}\right)\right)\right) \\ &- \sum_{j=1}^{N-1} \frac{\left(\Phi_j^{\mathrm{t}} - \Phi_{j-1}^{\mathrm{t}} - \Phi_j^{\mathrm{t,ext}} + \Phi_{j-1}^{\mathrm{t,ext}}\right)^2}{2L_{\mathrm{t}}^{\square}\frac{\Delta x}{w}} + \frac{\left(\Phi_j^{\mathrm{b}} - \Phi_{j-1}^{\mathrm{b}} - \Phi_j^{\mathrm{b,ext}} + \Phi_{j-1}^{\mathrm{b,ext}}\right)^2}{2L_{\mathrm{b}}^{\square}\frac{\Delta x}{w}}\end{aligned} \tag{A.1a}$$

$$\begin{aligned}= w \int_0^{\ell} \Bigg[&\frac{c_{\mathrm{J}}}{2} \left(\partial_t \Phi^{\mathrm{t}} - \partial_t \Phi^{\mathrm{b}}\right)^2 - \frac{\Phi_0 j_{\mathrm{c}}}{2\pi}\left(1 - \cos\left(\frac{2\pi}{\Phi_0}\left(\Phi^{\mathrm{t}} - \Phi^{\mathrm{b}}\right)\right)\right) \\ &- \frac{\left(\partial_x \Phi^{\mathrm{t}} - A^{\mathrm{t}}\right)^2}{2L_{\mathrm{t}}^{\square}} - \frac{\left(\partial_x \Phi^{\mathrm{b}} - A^{\mathrm{b}}\right)^2}{2L_{\mathrm{b}}^{\square}} \Bigg] \mathrm{d}x.\end{aligned} \tag{A.1b}$$

Here, $c_{\mathrm{J}} = C_{\mathrm{J}}^{\mathrm{tot}}/w\ell$ denotes the specific junction capacitance, $L_{\mathrm{t}}^{\square} = L_{\mathrm{t}}^{\mathrm{tot}} w/\ell$ and $L_{\mathrm{b}}^{\square} = L_{\mathrm{b}}^{\mathrm{tot}} w/\ell$ the top and bottom electrode's sheet inductance that sums up both the geometric and kinetic parts and $j_{\mathrm{c}} = I_{\mathrm{c}}^{\mathrm{tot}}/w\ell$ describe the critical and bias current density.

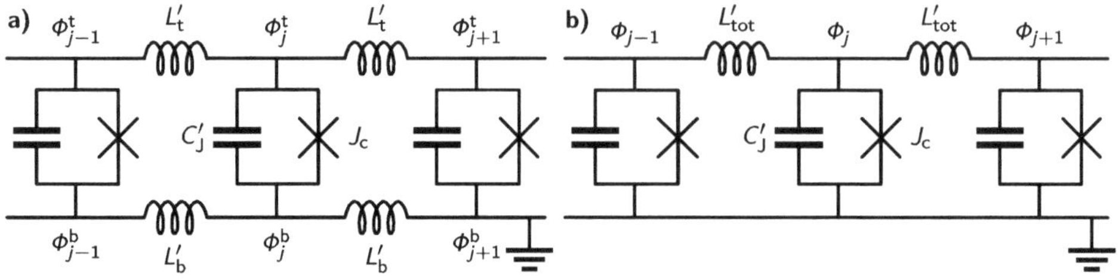

Figure A.1: Equivalent circuit of long Josephson junctions. a) A long junctions is modeled by lumped element junctions that are connected by inductive top and bottom electrodes, whose inductance may differ. **b)** In this circuit, the two inductance are equivalent to only one effective inductance in one electrode.

Applying the Euler-Lagrange equation [BP82; Leg87; CF91] yields the coupled equations of motion

$$0 = \frac{\mathrm{d}}{\mathrm{d}\bar{t}} \frac{\partial \bar{\mathcal{L}}}{\partial_{\bar{t}} \Phi^{\mathrm{t}\backslash\mathrm{b}}} + \frac{\mathrm{d}}{\mathrm{d}\bar{x}} \frac{\partial \bar{\mathcal{L}}}{\partial_{\bar{x}} \Phi^{\mathrm{t}\backslash\mathrm{b}}} - \frac{\partial \bar{\mathcal{L}}}{\partial \Phi^{\mathrm{t}\backslash\mathrm{b}}}, \tag{A.2a}$$

$$0 = c_{\mathrm{J}} \left(\partial_{tt} \Phi^{\mathrm{t}} - \partial_{tt} \Phi^{\mathrm{b}} \right) - \frac{\partial_{xx} \Phi^{\mathrm{t}} - A^{\mathrm{t}}}{L^{\square}_{\mathrm{t}}} + j_{\mathrm{c}} \sin\left(\frac{2\pi}{\Phi_0} \left(\Phi^{\mathrm{t}} - \Phi^{\mathrm{b}} \right) \right), \tag{A.2b}$$

$$0 = -c_{\mathrm{J}} \left(\partial_{tt} \Phi^{\mathrm{t}} - \partial_{tt} \Phi^{\mathrm{b}} \right) - \frac{\partial_{xx} \Phi^{\mathrm{b}} - A^{\mathrm{b}}}{L^{\square}_{\mathrm{t}}} - j_{\mathrm{c}} \sin\left(\frac{2\pi}{\Phi_0} \left(\Phi^{\mathrm{t}} - \Phi^{\mathrm{b}} \right) \right). \tag{A.2c}$$

Calculating (A.2b)/$L^{\square}_{\mathrm{b}}$ − (A.2c)/$L^{\square}_{\mathrm{t}}$ yields

$$0 = c_{\mathrm{J}} \left(\partial_{tt} \Phi^{\mathrm{t}} - \partial_{tt} \Phi^{\mathrm{b}} \right) - \frac{\partial_{xx} \Phi^{\mathrm{t}} - \partial_{xx} \Phi^{\mathrm{b}} - A^{\mathrm{t}} + A^{\mathrm{b}}}{L^{\square}_{\mathrm{t}} L^{\square}_{\mathrm{b}} \left(\frac{1}{L^{\square}_{\mathrm{b}}} + \frac{1}{L^{\square}_{\mathrm{t}}} \right)} - j_{\mathrm{c}} \sin\left(\frac{2\pi}{\Phi_0} \left(\Phi^{\mathrm{t}} - \Phi^{\mathrm{b}} \right) \right) \tag{A.3a}$$

$$= c_{\mathrm{J}} \partial_{tt} \Phi - \frac{\partial_{xx} \Phi - A}{L^{\square}_{\mathrm{tot}}} - j_{\mathrm{c}} \sin\left(\frac{2\pi}{\Phi_0} \Phi \right). \tag{A.3b}$$

In the last step, the node fluxes $\Phi^{\mathrm{t}} - \Phi^{\mathrm{b}} = \Phi$ are substituted by one single node flux, describing the difference of top and bottom electrode and the vector potentials $A^{\mathrm{t}} - A^{\mathrm{b}} = A$ are treated analogously. Furthermore, an effective electrode inductance is identified as $L^{\square}_{\mathrm{t}} L^{\square}_{\mathrm{b}} \left(\frac{1}{L^{\square}_{\mathrm{b}}} + \frac{1}{L^{\square}_{\mathrm{t}}} \right) = L^{\square}_{\mathrm{t}} + L^{\square}_{\mathrm{b}} = L^{\square}_{\mathrm{tot}}$ the sum of top and bottom electrode's inductance. Hence, the circuit can be simplified to having only one total inductance as shown in Fig. A.1 b).

A.2 Model of phase biased long Josephson junctions

In order to analyze the phase distribution of a distributed long junction is divided into $N \to \infty$ discrete segments and Kirchhoff's current law is applied at each node $j \in [1, N]$. In comparison to the high-kinetic inductance in the bottom electrode $L_0^{\mathrm{AlO}_x} \gg L_0^{\mathrm{Al}}$, the top electrode's purely geometric inductance L_0^{Al} is negligible, and the jth node flux is coupled by the nearest neighbors as

$$\frac{\Phi_{j+1} - \Phi_j - \Phi_{j+1}^{\mathrm{ext}} + \Phi_j^{\mathrm{ext}}}{L_0/(N-1)} + \frac{I_\mathrm{c}}{N} \sin\left(\frac{2\pi}{\Phi_0}\Phi_j\right) = \frac{\Phi_j - \Phi_{j-1} - \Phi_j^{\mathrm{ext}} + \Phi_{j-1}^{\mathrm{ext}}}{L_0/(N-1)}, \tag{A.4a}$$

$$\frac{\Phi_2 - \Phi_1 - \Phi_2^{\mathrm{ext}} + \Phi_1^{\mathrm{ext}}}{L_0/(N-1)} - \frac{I_\mathrm{c}}{N} \sin\left(\frac{2\pi}{\Phi_0}\Phi_1\right) = I, \tag{A.4b}$$

$$I - \frac{I_\mathrm{c}}{N} \sin\left(\frac{2\pi}{\Phi_0}\Phi_{N-1}\right) = \frac{\Phi_N - \Phi_{N-1} - \Phi_N^{\mathrm{ext}} + \Phi_{N-1}^{\mathrm{ext}}}{L_0/(N-1)}, \tag{A.4c}$$

including the boundary conditions at the junction edges, defined by the bias current I. The node fluxes can be translated into node phases by means of the junction's well-known flux-phase relation $\Phi_j = \frac{\Phi_0}{2\pi}\varphi_j$. Besides, a homogeneous external magnetic flux causes a constant external phase drop between two adjacent nodes $\Phi_{j+1}^{\mathrm{ext}} - \Phi_j^{\mathrm{ext}} = \frac{\Phi_0}{2\pi}\frac{\varphi^{\mathrm{ext}}}{N-1} = \frac{\Phi_0}{2\pi}\Delta\varphi^{\mathrm{ext}}\ \forall\, j$.

$$\frac{\varphi_{j+1} - \varphi_j - \varphi_{j+1}^{\mathrm{ext}} + \varphi_j^{\mathrm{ext}}}{L_0/(N-1)} + \frac{2\pi I_\mathrm{c}}{N\Phi_0}\sin(\varphi_j) = \frac{\varphi_j - \varphi_{j-1} - \varphi_j^{\mathrm{ext}} + \varphi_{j-1}^{\mathrm{ext}}}{L_0/(N-1)}, \tag{A.5a}$$

$$\frac{\varphi_2 - \varphi_1 - \varphi_2^{\mathrm{ext}} + \varphi_1^{\mathrm{ext}}}{L_0/(N-1)} - \frac{2\pi I_\mathrm{c}}{N\Phi_0}\sin(\varphi_1) = I, \tag{A.5b}$$

$$I - \frac{2\pi I_\mathrm{c}}{N\Phi_0}\sin(\varphi_{N-1}) = \frac{\varphi_N - \varphi_{N-1} - \varphi_N^{\mathrm{ext}} + \varphi_{N-1}^{\mathrm{ext}}}{L_0/(N-1)}. \tag{A.5c}$$

By further identifying the Josephson inductance $L_\mathrm{J} = \frac{\Phi_0}{2\pi I_\mathrm{c}}$ and introducing the screening parameter $\beta_\mathrm{L} = \frac{L_0}{L_\mathrm{J}} = \frac{2\pi I_\mathrm{c} L_0}{\Phi_0}$, the conditional equations result in

$$\varphi_2 = \varphi_1 + \Delta\varphi^{\mathrm{ext}} + \frac{\beta_\mathrm{L}}{N(N-1)}\sin(\varphi_0) + \frac{\beta_\mathrm{L}\gamma}{N-1}, \tag{A.6a}$$

$$\varphi_{j+1} = 2\varphi_j - \varphi_{j-1} - \frac{\beta_\mathrm{L}}{N(N-1)}\sin(\varphi_j), \tag{A.6b}$$

$$\frac{\beta_\mathrm{L}\gamma}{N-1} = \varphi_N - \varphi_{N-1} - \Delta\varphi^{\mathrm{ext}} + \frac{\beta_\mathrm{L}}{N(N-1)}\sin(\varphi_N). \tag{A.6c}$$

B Experimental details

B.1 Cryogenic measurement setup

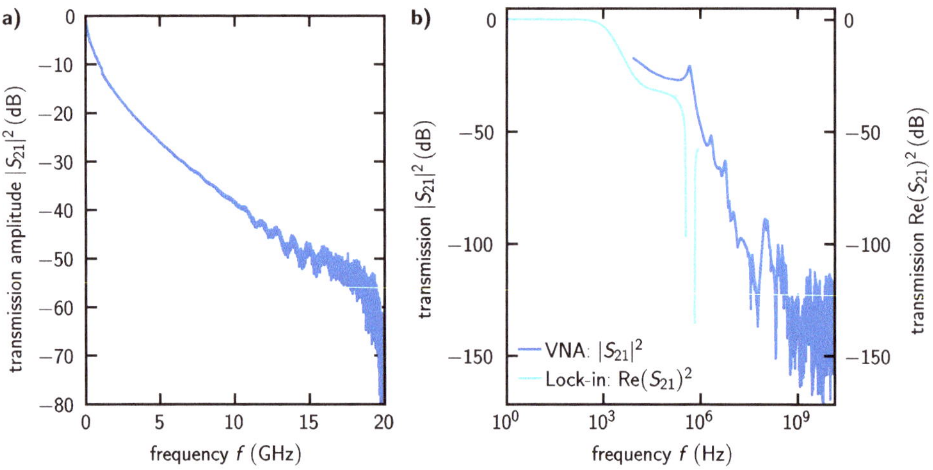

Figure B.1: Spectra of RF and DC measurement lines. a) Spectrum of RF signal lines at room temperature. The attenuation of the stainless steel coaxial cables shows the typical $1/\sqrt{f}$ behavior if the wavelength λ is smaller than the diameter of the coaxial line $\approx 0.3\,\mathrm{mm}$. **b)** Spectrum of DC measurement lines at cryogenic temperatures. The spectrum shows characteristics of two low-pass filters with $\approx 30\,\mathrm{dB}$ attenuation each. The lower cut-off frequency $\approx 1\,\mathrm{kHz}$ is attributed to a *RCR* filter at the second pulse tube stage, made from temperature insensitive thin-film resistors with $R = 180\,\Omega$ and multilayer ceramic (NP0) capacitors with $C = 1\,\mathrm{nF}$. The upper cut-off frequency 300 kHz corresponds to the *CLC* low-pass filter at room temperature, consisting of shielded wire-wound SMD inductor with a ferrite core with $L = 100\,\mu\mathrm{H}$ and multilayer ceramic (NP0) capacitors with $C = 1\,\mathrm{nF}$. This type of *LC* circuit shows a different spectrum in resonance depending on whether the complex or real-valued signal is measured. The data in the low-frequency range from 1 Hz to 5 MHz are acquired with a Lock-in amplifier, whereas the microwave range from 14 kHz to 20 GHz is measured with a VNA.

B.2 Fabrication parameters

Table B.1: Resist application. To coat a sample with a resist of thickness d, the spin coater accelerates with a, where the resist is distributed of the whole chip in a first ramping step with speed v_r for the duration t_r. In a second step, the targeted thickness is achieved by spinning with speed v_s for the time t_s. Afterwards the resist is softbaked at the temperature T_B for the time t_B to vapor solvents and set the resist.

resist	d (μm)	a (rpm/s)	v_r (rpm)	t_r (s)	v_s (rpm)	t_s (s)	T_B (°C)	t_B (s)
S1805 (optical)	~0.5	6000	500	1	6000	60	115	60
AZ5214E (optical)	~1.2	7500	500	5	6000	60	110	50
ma-N 2403 (e-beam)	~0.3	1000			3000	65	90	60
S1818 (protective)	~1.8	2000			4500	60	80	300

Table B.2: Optical resist exposure parameters. A XeHg lamp emitting the wavelength $\lambda = 365\,\mathrm{nm}$ is used to expose the photoresist under a mask for the time t_{ex}. Thereby a constant power mode, controlling the power P_{ex}, or constant intensity mode, adjusting the dose D_{ex}, is available.

resist	process	mask	P_{ex} (W)	D_{ex} (mW cm^{-2})	t_{ex} (s)
S1805	positive	soda lime	370		1.8
AZ5214E	positive	soda lime		13.0	5.0

Table B.3: Electron-beam resist exposure parameters. Depending on the required resolution and accuracy of the desired structures, two different beam currents I, step sizes Δx are used to pattern the ma-N 2403 resist, whereas the average area dose D is kept constant. The small beam current and step size is used for writing the 200 nm wide long junction structures. The microstrip feedline and DC leads of test structures, however, are written with higher beam currents and step sizes in order to save writing time.

resist	D (μC cm^{-2})	I (nA)	Δx (nm)
ma-N 2403	240	2	10
ma-N 2403	240	30	50

Table B.4: Resist development parameters. Each exposed resist is developed in a selective solvent (developer) for a time t_d, which is stopped in a stop bath.

resist	developer	t_d (s)	stop bath
S1805	AZ developer : H_2O (3 : 2)	33	H_2O
AZ5214E	AZ developer : H_2O (1 : 1)	33	H_2O
ma-N 2403	MF-319	50	H_2O

Table B.5: ICP trilayer etching parameters. Depending on requirements of (an-) isotropy and etching rate, different ICP processes are used that differ in participating gases, RF, and ICP power. The etching time is chosen to etch the desired layers as properly as possible.

trilayer	layers	I_{Ar} (sccm)	I_{Cl_2} (sccm)	I_{O_2} (sccm)	P_{RF} (W)	P_{ICP} (W)	t (s)
A	trilayer	15	15	0	50	100	205
A	top	15	15	0	50	100	90
B	trilayer	15	15	0	50	100	180
B	top	15	15	0	50	100	90
C	trilayer	15	15	0	50	100	215
C	top	15	15	0	50	100	110
D	trilayer	2	12	0	100	200	70
D	top	15	3	1	100	200	32
E	top	15	3	1	100	100	5
E	bottom	2	12	0	100	200	120

Table B.6: Anodic oxidation parameters. The sample is facing a platinum anode in a fixed distance of 2 cm in an electrolyte of 156 g ammonium pentaborate in 1120 ml ethylene glycol and 760 ml demineralized water. The oxide layer is defined by the applied voltage V and the measured current saturated over the time t. The influence of different voltages was studied at trilayer A, where three 5 mm × 5 mm in a row were dipped successive deeper and deeper in the solution by decreasing the voltage.

trilayer	chip	V (V)	t (min)
A	1–3	9	15
A	4–6	11	15
A	7–9	13	15
B	1–9	11	15
C	1–9	11	15

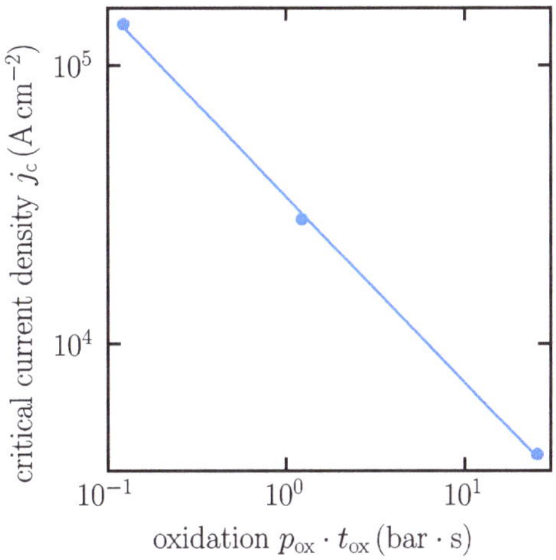

Figure B.2: Static oxidation parameters for tunnel barrier creation. The critical current densities of trilayers A, B, and C are determined from short squared junctions measured at millikelvin temperatures. According to Kleinsasser *et al.* [KMM95], j_c depends exponentially on the product of oxygen (partial) pressure p_{ox} and oxidation time t_{ox}.

C Additional measurement data

Table C.1: Design parameters of Trilayer D for long junctions, each embedded in a resonator. A long junction of length ℓ and width w is placed in the center of a rectangular microstrip resonator of length l and width w.

device	l (µm)	w (µm)	ℓ (µm)	w (µm)
Res #1	503.0	2.0	50.3	2.0
Res #2	440.4	2.0	44.0	2.0
Res #3	390.9	2.0	0.0	0.0
Res #4	350.8	2.0	35.1	2.0
Res #5	317.7	2.0	31.8	2.0
Res #6	289.9	2.0	29.0	2.0

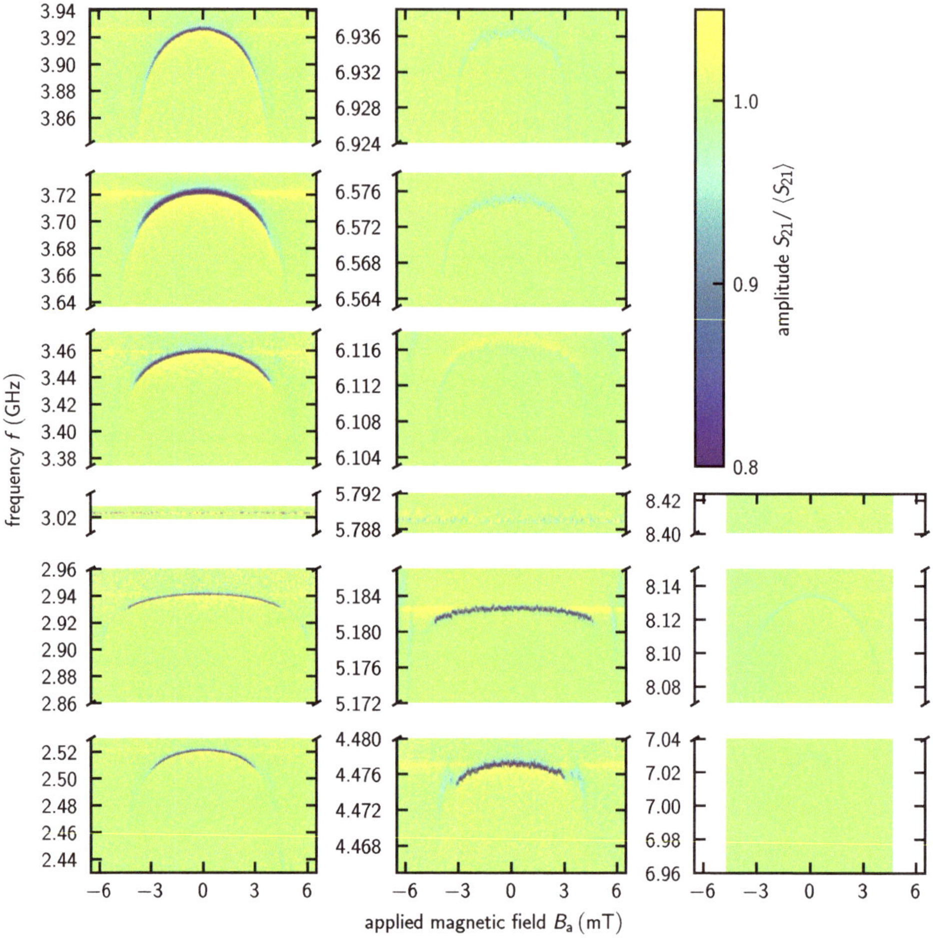

Figure C.1: Magnetic field dependence of coupled systems of a long junction embedded in a resonator. The amplitude's frequency spectrum $|S_{21}|$ of Res #1 in the lowest row to the Res #6 in the uppermost row is depicted as color depending on the applied magnetic field. The magnetic field dependence of the system's resonant frequency (dark lines) differs significantly for the first (left), second (center), and third (right) harmonics caused by the LJJ's magnetic field dependent inductance. (Note the different frequency scales for different harmonics in the three columns.) The eigenfrequencies of the reference resonator (Res #3) feature no magnetic field dependence.

Acknowledgments

Ich bin sehr dankbar, dass ich all die Anstrengungen dieser Doktorarbeit nicht alleine bewältigen musste und ich von vielen Menschen umgeben war, die mich auf diesem Weg unterstützt haben, von denen ich viel lernen konnte und die mir mit Rat und Tat zur Seite gestanden sind. An dieser Stelle möchte ich mich bei euch allen bedanken, die ihr auf unterschiedlichste Weise zum Gelingen dieser Arbeit beigetragen und sie damit ermöglicht habt.

Zuallererst bedanke ich mich vielmals bei Prof. Alexey Ustinov, von dessen Gruppe ich ab meiner Bachelorarbeit Teil sein konnte und der auch diese Arbeit ermöglicht und betreut hat. Vielen Dank für das großartige Promotionsthema und dass wir gemeinsam die Fluxonen in unserer "fluxon group" wieder aufleben lassen konnten. Ich schätze besonders die hilfreichen Diskussionen, deine große physikalische Erfahrung und dein breites Wissen auf dem Gebiet der Josephsonkontakte.

Vielen Dank an Prof. Alexander Shnirman für die Zweitbetreuung der vorliegenden Arbeit und die wertvollen Diskussionen über die theoretischen Grundlagen als Vorüberlegungen für meine Experimente.

Außerordentlicher Dank gilt Dr. Hannes Rotzinger, der die größte Hilfe bei der tagtäglichen Arbeit im Labor war. Dankeschön, dass du immer ein offenes Ohr hast, ich dich alles fragen kann und du dabei auch über den Tellerrand hinausblickst. In den schwierigeren Phasen konnte ich mir sicher sein, dass du mich unterstützt, mir etliche neue Impulse und kreative Ideen zum Weiterarbeiten gibst. Am allermeisten schätze ich aber, dass du nicht nur fachlich, sondern auch menschlich ein toller Chef bist, der mich bei Unsicherheiten unterstützt und aufbaut und mir notfalls auch ein Päckchen zur körperlichen Stärkung nach Hause schickt.

Herzlichen Dank an Dr. Andre Schneider, dessen große Hilfsbereitschaft mir sehr wichtig war und die ich bewundere. Danke, dass du mich ab unserem ersten Tag in der Gruppe begleitet hast und mich mit deinem Wissen über supraleitende Qubits, deiner Erfahrung bei Time-Domain Messungen und bei jeglichen Programmierproblemen unterstützt hast. Du warst mir durch die Begleitung und das Korrekturlesen meiner Arbeiten eine enorme Hilfe, ein guter Freund und für mich mit deiner Art die gute Seele der Arbeitsgruppe.

Mein großer Dank gilt auch Dr. Jürgen Lisenfeld dafür, dass du deine enorme experimentelle Erfahrung mit mir geteilt hast, all meine Pläne kritisch hinterfragt hast und mir damit so manchen Iterationsschritt erspart hast.

Vielen Dank Dr. Mikhail Fistul für allen theoretischen Input und das Korrekturlesen, was maßgeblich das Fortschreiten meiner Arbeit ermöglichte.

Ich bedanke mich aus tiefstem Herzen bei Martin Spiecker für deine große Hilfsbereitschaft, das Teilen deines riesigen Wissens über die Quantenmechanik und deine wertvollen Gespräche. Danke, dass wir über die unterschiedlichen Arbeitsgruppen hinweg so gut miteinander arbeiten konnten, du die Arbeitsatmosphäre bereichert hast und du mir so zu einem guten Freund geworden bist.

Tim Wolz, dir gebührt ein besonders großer Dank für deine große Hilfe beim Schreiben und deiner Mühe, denselben Text immer wieder zu lesen und zu verbessern. Es war mir eine Freude, dass du mich in den letzten Jahren nicht nur als Kollege, sondern auch als Freund und WG-Mitbewohner begleitet hast und wir so selbst im Lockdown Home-Office noch über unsere tagtägliche Arbeit und auch sonst über Gott und die Welt sprechen konnten.

Mein besonderer Dank gilt allen Studenten, die ich betreuen durfte und die alle ihren Teil zum Fortschritt meines Projekts beigetragen haben. Vielen Dank Lukas Powalla für deine Hilfe bei den ersten gemeinsamen Schritten der Fabrikation, deine Ausdauer und deinen Ehrgeiz, durch die wir zusammen so manche Nacht im Labor verbracht haben. Danke Konrad Dapper für all deine theoretischen Überlegungen und deine Hilfe bei numerischen Simulationen.

Ich bin dankbar und glücklich, die ganze Zeit über so nette und clevere Kollegen um mich gehabt zu haben, die eine angenehme und produktive Atmosphäre geschaffen haben. Insbesondere danke ich Jan Nicolas Voss, Maximilian Kristen, Yannick Schön, Alexander Stehli, Jan Brehm, Alexander Bilmes und Alex Kreuzer für die Zusammenarbeit und eure Unterstützung bei den tagtäglichen Problemen im Labor und für den angenehmen Umgang untereinander. Besonderer Dank gilt Konstantin Händel, András Di Giovanni und Hossam Tohamy für das sorgfältige Korrekturlesen dieser Arbeit.

Ich bedanke mich von Herzen bei Lucas Radtke für seinen unermüdlichen Einsatz im Reinraum, seine ständige Hilfsbereitschaft und das Sägen unzähliger Wafer. Danke Dr. Silvia Diewald für den Einsatz der Elektronenstrahllithografie und Danke Dr. Aina Quintilla für deine Notfallhilfe, wenn die ICP-Anlage zum wiederholten Mal gestreikt hat und man dich auch nach Feierabend noch anrufen konnte, um Proben rechtzeitig fertig zu bekommen.

Vielen Dank an die Werkstätten des Physikalischen Instituts, stellvertretend Michael Meyer, Patrick Rust und Jannis Ret. Danke für all die gefertigten Werkstücke und für all die Tipps, die ihr mir gegeben habt, durch die ich viel über Metallbearbeitung und Elektrotechnik lernen konnte.

Ein Dankeschön gilt Steffi Baatz und Claudia Alaya für die unkomplizierte Hilfe bei allen administrativen Themen, wie beispielsweise den unzähligen Arbeitsverträgen, um uns Wissenschaftlern den Rücken freizuhalten.

An dieser Stelle möchte ich mich auch bei der Helmholtz International Research School for Teratronics und der Landesgraduiertenförderung Baden-Württemberg für die finanzielle Unterstützung dieser Promotion bedanken.

Der größte Dank gilt meiner Familie und meinen Freunden: Danke Steffi für alle aufmunternden und motivierende Worte, für alle Lebensfreude, die du vermittelst, für dein Verständnis für lange, harte Arbeit und dass du es ertragen hast, viele Stunden auf mich zu verzichten. Danke, dass du mir auch in stressigen Zeiten zeigst, dass es im Leben so viel mehr als Physik gibt, mich auf andere Gedanken bringst und dass du mich so nimmst, wie ich bin. Ich liebe dich und bin unendlich dankbar, dich an meiner Seite zu haben. Danke Mama und Papa, dass ihr mich mein ganzes Leben lang so liebevoll begleitet habt, mir unendlich viele Dinge beigebracht habt und mir das Physikstudium ermöglicht habt. Ohne euch wäre ich nicht der, der ich heute bin und hätte diese Arbeit nicht machen können. Danke Sarah mit Familie, dass ich immer zu euch kommen kann, ihr immer für mich da seid und ihr mich mit allem euch Möglichem unterstützt.

Karlsruhe, Juli 2022 Micha Wildermuth

Experimental Condensed Matter Physics
(ISSN 2191-9925)

Band 1 Alexey Feofanov
Experiments on flux qubits with pi-shifters.
ISBN 978-3-86644-644-1

Band 2 Stefan Schmaus
Spintronics with individual metal-organic molecules.
ISBN 978-3-86644-649-6

Band 3 Marc Müller
Elektrischer Leitwert von magnetostriktiven Dy-Nanokontakten.
ISBN 978-3-86644-726-4

Band 4 Torben Peichl
Einfluss mechanischer Deformation auf atomare Tunnelsysteme – untersucht mit Josephson Phasen-Qubits.
ISBN 978-3-86644-837-7

Band 5 Dominik Stöffler
Herstellung dünner metallischer Brücken durch Elektromigration und Charakterisierung mit Rastersondentechniken.
ISBN 978-3-86644-843-8

Band 6 Tihomir Tomanic
Untersuchung des elektronischen Oberflächenzustands von Ag-Inseln auf supraleitendem Niob (110).
ISBN 978-3-86644-898-8

Band 7 Lukas Gerhard
Magnetoelectric coupling at metal surfaces.
ISBN 978-3-7315-0063-6

Band 8 Nicht erschienen.

Band 9 Jochen Zimmer
Cooper pair transport in arrays of Josephson junctions.
ISBN 978-3-7315-0130-5

Band 10 Oliver Berg
Elektrischer Transport durch Nanokontakte von Selten-Erd-Metallen.
ISBN 978-3-7315-0209-8

Band 11 Grigorij Jur'evic Grabovskij
Investigation of coherent microscopic defects inside the tunneling barrier of a Josephson junction.
ISBN 978-3-7315-0210-4

Die Bände sind unter www.ksp.kit.edu als PDF frei verfügbar oder als Druckausgabe bestellbar.

Band 12	Cornelius Thiele **STM Characterization of Phenylene-Ethynylene Oligomers on Au(111) and their Integration into Carbon Nanotube Nanogaps.** ISBN 978-3-7315-0235-7
Band 13	Michael Peter Schackert **Scanning Tunneling Spectroscopy on Electron-Boson Interactions in Superconductors.** ISBN 978-3-7315-0238-8
Band 14	Susanne Butz **One-Dimensional Tunable Josephson Metamaterials.** ISBN 978-3-7315-0271-5
Band 15	Philipp Jung **Nonlinear Effects in Superconducting Quantum Interference Meta-Atoms.** ISBN 978-3-7315-0294-4
Band 16	Sebastian Probst **Hybrid quantum system based on rare earth doped crystals.** ISBN 978-3-7315-0345-3
Band 17	Wolfram Kittler **Magnetische Anisotropie und Quantenphasenübergang in $CeTi_{1-x}V_xGe_3$.** ISBN 978-3-7315-0363-7
Band 18	Moritz Peter **Towards magnetic resonance in scanning tunneling microscopy using heterodyne detection.** ISBN 978-3-7315-0410-8
Band 19	Junji Tobias Märkl **Investigation of Magnetic Adatoms with Scanning Tunneling Techniques.** ISBN 978-3-7315-0435-1
Band 20	Jochen Braumüller **Quantum simulation experiments with superconducting circuits.** ISBN 978-3-7315-0780-2
Band 21	Jinjie Chen **Local Investigation of Single Magnetic Molecules with Scanning Tunneling Microscopy.** ISBN 978-3-7315-0819-9
Band 22	Arnold Seiler **Einfluss der Leitungselektronen auf die Dynamik atomarer Tunnelsysteme in ungeordneten Festkörpern: Relaxationsprozesse in metallischen Gläsern und ungeordneten dünnen Aluminiumoxid-Schichten.** ISBN 978-3-7315-0870-0

Die Bände sind unter www.ksp.kit.edu als PDF frei verfügbar oder als Druckausgabe bestellbar.

Band 23 Jasmin Maria Jandke
Elastic and Inelastic Scanning Tunneling Spectroscopy on Iron-Based Superconductors.
ISBN 978-3-7315-0747-5

Band 24 Kevin Edelmann
Electroluminescence from Plasmonic Excitations in a Scanning Tunnelling Microscope.
ISBN 978-3-7315-0923-3

Band 25 Lei Zhang
Sub-Kelvin scanning tunneling microscopy on magnetic molecules.
ISBN 978-3-86644-950-3

Band 26 Lukas Grünhaupt
Granular aluminium superinductors.
ISBN 978-3-7315-0956-1

Band 27 Alexander Bilmes
Resolving locations of defects in superconducting transmon qubits.
ISBN 978-3-7315-0967-7

Band 28 Marco Pfirrmann
Adding nonlinearity to an electromagnetic-magnonic quantum hybrid device.
ISBN 978-3-7315-1003-1

Band 29 Andre Schneider
Quantum Sensing Experiments with Superconducting Qubits.
ISBN 978-3-7315-1032-1

Band 30 Vibhuti Narayan Rai
Light Emission from Single Self-decoupled Molecules in a Scanning Tunnelling Microscope.
ISBN 978-3-7315-1201-1

Band 31 Julian Antoni Skolaut
Molecular Motor Based on Single Chiral Tripodal Molecules Studied with STM.
ISBN 978-3-7315-1247-9

Band 32 Thomas Gozlinski
Local Probing of a Superconductor's Quasiparticles and Bosonic Excitations with a Scanning Tunnelling Microscope.
ISBN 978-3-7315-1279-0

Band 33 Matthias Micha Wildermuth
Quantum Tunneling of Josephson Vortices in High-Impedance Long Junctions.
ISBN 978-3-7315-1303-2